STRENGTHENING RESEARCH CAPACITY AND DISSEMINATING NEW FINDINGS IN NURSING AND PUBLIC HEALTH

PROCEEDINGS OF THE 1ST ANDALAS INTERNATIONAL NURSING CONFERENCE (AINiC 2017), PADANG, INDONESIA, 25–27 SEPTEMBER 2017

# Strengthening Research Capacity and Disseminating New Findings in Nursing and Public Health

*Editors*

Hema Malini

*Faculty of Nursing, University of Andalas, Padang, Indonesia*

Khatijah Lim Abdullah

*Department of Nursing Sciences, Faculty of Medicine, University of Malaya, Kuala Lumpur, Malaysia*

Judith McFarlane

*Nelda C. Stark College of Nursing, Texas Woman's University, Houston, USA*

Jeff Evans

*Faculty of Life Sciences and Education, University of South Wales, UK*

Yanti Puspita Sari

*Faculty of Nursing, University of Andalas, Padang, Indonesia*

CRC Press
Taylor & Francis Group
Boca Raton London New York Leiden

CRC Press is an imprint of the
Taylor & Francis Group, an **informa** business
A BALKEMA BOOK

Cover illustration: Jam Gadang, the clock tower at Bukittinggi in West Sumatra, Indonesia.
Photographer: Ikhwan Arief.

*CRC Press/Balkema is an imprint of the Taylor & Francis Group, an informa business*



Typeset by V Publishing Solutions Pvt Ltd., Chennai, India



Published by: CRC Press/Balkema
Schipholweg 107C, 2316 XC Leiden, The Netherlands
e-mail: Pub.NL@taylorandfrancis.com
www.crcpress.com – www.taylorandfrancis.com

ISBN: 978-1-138-50066-2 (Hbk + CD-ROM)
ISBN: 978-1-315-14390-3 (eBook)

*Strengthening Research Capacity and Disseminating New Findings in Nursing and Public Health – Malini et al. (Eds)*
*© 2018 Taylor & Francis Group, London, ISBN 978-1-138-50066-2*

# Table of contents

*Strengthening Research Capacity and Disseminating New Findings in Nursing and Public Health – Malini et al. (Eds)*
*© 2018 Taylor & Francis Group, London, ISBN 978-1-138-50066-2*

# Preface

The proceedings of the 1st Andalas International Nursing Conference (AINiC) 2017 is a compilation of the papers presented at the conference held in Padang, West Sumatra, Indonesia, 25–27 September 2017. The conference has brought together a variety of professionals and researchers in the field of nursing science and education; e.g. nursing experts, nursing researchers, healthcare professionals, nursing educators, and students. This book contains keynote speeches, plenary sessions and research presentations.

We expect the proceedings will contribute to research and development in learning, the disseminating of new findings in nursing and to stimulate networking within nursing practice, nursing research and nursing education.

AINiC 2017 Committee

*Strengthening Research Capacity and Disseminating New Findings in Nursing and Public Health – Malini et al. (Eds)*
*© 2018 Taylor & Francis Group, London, ISBN 978-1-138-50066-2*

# Foreword

Nursing practice and research have been continuously developed to meet new challenges. Research and practice development are important to inititate new strategies to provide alternatives and to face the needs of related elements in educating nurses better for the future.

The present proceedings comprise the selection of contributed papers of the 1st Andalas International Nursing Conference (AINiC 2017) in September 2017 by the University of Andalas. The Conference provided an opportunity to present newly developed and/or applied practice approaches and new strategies in nursing practice and service development. It was also an important meeting place for researchers and educators providing opportunities to meet each other and establishing new network facilities.

The seminar was structured in plenary sessions, oral sessions and poster sessions. Papers presented covered a wide range of topics, regional and national issues. We wish to thank all participants, and presenters in particular, for their efforts and contributions to the success of the seminar. We thankfully acknowledge the valuable comments of the reviewers to these seminar proceedings.

Editors

*Strengthening Research Capacity and Disseminating New Findings in Nursing and Public Health – Malini et al. (Eds)*
*© 2018 Taylor & Francis Group, London, ISBN 978-1-138-50066-2*

# Conference objectives

- Accommodate and increase research contributions in the field of nursing and health
- Facilitate research publication
- Improve the understanding and mastery of nursing and health
- Initiate the possibility of establishing cooperation amongst individuals and institutions

*Strengthening Research Capacity and Disseminating New Findings in Nursing and Public Health – Malini et al. (Eds)*
*© 2018 Taylor & Francis Group, London, ISBN 978-1-138-50066-2*

# Conference program

## MONDAY 25 SEPTEMBER 2017

Opening Ceremony and Welcome Speech

*Cultural tribute and art*

Tari Galombang (Traditional Dances)
Video Playing
Indonesia Raya National Anthem
Holy Quran

## WELCOME SPEECHES

AINiC Committee President Welcome Speech
Faculty of Nursing Dean Speech
University of Andalas Rector Speech
Governor of West Sumatra Welcome Speech
Directorat General Dikti Speech
Traditional Dance (Piring Dance)
Session praying

*Strengthening Research Capacity and Disseminating New Findings in Nursing and Public Health – Malini et al. (Eds)*
*© 2018 Taylor & Francis Group, London, ISBN 978-1-138-50066-2*

# Plenary speech

**Distinguished speakers and participants**

It is my great pleasure to be here today to address this first Andalas International Nursing Conference (AINiC 2017).

We are very grateful to welcome you all in Padang City. We hope you all enjoy the city as you enjoy the conference.

Welcome to the motherland of victorious Minangkabau.

We are fully fortunate to have the support from all of the important institutions: the city government, university, invited speakers and sponsors.

I would like to refer much gratitude and a warm welcome to the Directorate General Higher Education of Indonesia, our fully respected Governor West Sumatra, and our University of Andalas Rector.

Please allow me to send an appreciation to all members of the organizing committee who have put together and carried out the program.

Research has clearly brought into focus the need for literature development and to guide clinical practice decisions. Engaging in research is a way to influence practice and policy, to learn about research itself, to learn about oneself, and to change the way we think. Being actively involved in research will provide opportunities to gain a more in-depth understanding of knowledge and a renewed perspective.

In recent years, health research publications have been highlighted as important to science knowledge development and its strong relationship to clinical practice and social development elements. In nursing itself, we know the term Evidence Based Nursing Practice (EBNP) which means nursing professionals employ the results of research in their daily intervention nursing care.

Publication is important as a key way for professionals and researchers to validate their work, and in the wider scheme of matters, to create novel solutions to complex problems through dialogue with fellow researchers. Publishing research is important because we all benefit from cooperation. Publication is the way that scientists communicate their findings with others. And the larger the cooperating group, the better. Publication lets the collaborating group expand to include the whole world and people who will enter our field in the future.

Publishing research will boost scientists' recognition, generate invitations to meetings, present consulting and collaboration opportunities, and increases citation rates because their productivity will be more visible. Making research to be published maximizes the potential return on the investment in research, and those researches can be repatriated to the countries from which they may have been collected by foreign scientists.

Research publishing will give people the opportunity to read our research results, they may build on our work, repeating, correcting, extending it to new cases, applying methods, results and the conclusions in ways we never thought of. In exchange, other publications may benefit us as well.

The primary motivations for us as scientists to publish are to demonstrate our contribution to science, and the consequent peer-recognition that influences one's reputation and employment opportunities, promotion at work, and ability to win further research funding.

Personal satisfaction in completing a study and enthusiasm about communicating findings and opinions to society. How does it make you satisfied! You all know the feeling.

One of the key elements of our meeting today is to minimize the gap between research and research publication willingness in order to increase research participation and research implementation.

It is such a great opportunity we gather here together to boost research impact.

Welcome and Enjoy yourself in Padang!

DR. MOHAMMAD DIMYATI

Research and Development Directorate General Ministry of Research, Technology and Higher Education

"Health Research Trend and Focus: How to Win International and National Research Funding and Grants"

The research funding comes from several resources. It is also common knowledge that funding is granted based on a competitive criteria. Research financial funding matters in every sector, including medical and health science. Research funding will provide the right circumstances to ensure the sustainability of research continuation and impact. There are rigorous requirements to win research grants.

Among the highlighted topics in the general plenary session:

- International grants opportunities
- National grants opportunities
- Tips to win and proposal preparation

PROF. DR. RIZANDA MACHMUD, MD, FISPH, FISCM

"Mental Health Focus is Now Global Public Health Challenges"

The Movement for Global Mental Health is a network of individuals and organizations that aim to improve services for people living with mental health problems and psychosocial disabilities worldwide, especially in low- and middle-income countries where effective services are often scarce. Two principles are fundamental to the Movement: scientific evidence and human rights.

Among the highlighted topics in the public health plenary session:

- Global perspective of mental health into public health responsibilities
- Cultural and contextual challenges of mental health concerns in public health
- Policy development in mental health as public health focus

PROF. LISA McKENNA, PHD, RN RM FACN

"Preparing Qualified Nurses: Nursing Education Institution Major Role in Providing Excellence Standard of Nursing Education, Graduate Transition and Work Readiness."

Nursing education in Indonesia is currently facing challenges related to the application of nursing knowledge in clinical environments and inability of students in application of nursing procedures in clinical settings. Nursing education institutions need to pay careful attention to educational performance and clinical skills, as well as to determination of standards and validation of education quality.

Nursing education shall fulfill the competency demanded. This shift process should make a combination of serious hard work of the educational institutions and practical field institutions to provide supports to meet the achievement of expected competencies.

Among the highlights topic in nursing education plenary session:

- Nursing education global standard
- Nursing education institution strategy to enhance quality
- Graduate transition to work readiness: shorten waiting and wasting latent time

PROF. DR. KHATIJAH BINTI ABDULLAH LIM GEOK KHIM

"Theoretical development of Nursing "Where are we now"; A Recent Update of Theoretical Foundation in Nursing."

The development of knowledge for nursing poses an exciting, scholarly adventure for the profession's scientists. A series of challenges are involved: the challenge to develop the substantive content needed for practice within nursing's disciplinary perspective, the challenge to sustain excellence in the developing science base and in the preparation of nurse researchers, and the challenge of disseminating stable, appropriate research results to the profession's clinicians and to the public.

Nursing is entering a new era, moving from the stage of establishing structures to support nursing research and building the cadre of scientists needed to conduct investigations, to the stage of focusing on the identification and study of the phenomena which comprise the body of knowledge needed for practice. A number of directions or priorities for nursing research are evident for the future. Research based practice requires a merger of the talents and expertise of those providing practice and those developing the knowledge base for the profession.

Among the highlights topic in nursing science plenary session:

- Brief explanation of the progress of theoretical past-present in nursing
- Theoretical focus on development in nursing
- Nursing theoretical update and trend; new model and theory in nursing

NS. YUFITRIANA AMIR, S.KEP, MSC, PHD., FISQUA

"Repealing Health Care Inequality to Achieve Health Care Quality Improvements: Technical and Policy Advice."

AINiC 2017 will be the place for nursing professionals and public health implementation. Health inequality on the other hand designates differences, variations and disparities in health achievements of individuals and groups. Health equality does not imply moral judgement. The crux of the distinction between equality and equity is that the identification of health inequities entails normative judgement premised upon one's concept of justice, society and reasoning underlying the genesis of health inequalities.

Inequalities in health between population groups exist in all countries. These differences occur along several axes of social stratification including socioeconomic, political, ethnic, cultural and gender. The causes of inequalities in developed may be different from those in developing countries. However, in developing countries improved health among the urban population has been found to be due to access to improved health care knowledge and services.

Among the highlights topic in healthcare service plenary session:

- Healthcare inequality addressing socioeconomic and racial disparities
- Healthcare financing impact inequality
- Policy development in improving healthcare quality

HEMA MALINI, SKP., MN., PHD

"Challenges and Strategies in Conducting Culturally Relevant Health Education Program in Indonesia"

Providing health education to people with chronic disease such as diabetes becomes an important aspect in order to improve the quality of life. However, the structured education program is rarely implemented since there is limited information about how the education program should be conducted. The aim of this paper is to discuss several aspects that has been identified during the implementation and evaluation of a proposed structured education program named InGDEP in Indonesian setting. Several aspects of health beliefs and values of people with chronic diseases, such as the beliefs of traditional medicine and religious aspects, have been identified. The perception of health professionals concerning the implementation of structured education program is also analyzed. During the program implementation, there were some challenges and strategies that have been applied and proven to be effective during the program. It is recommended that the structured education program should become one alternative in delivering health information to people with chronic disease in Indonesia.

# *Keynote speakers*

*Strengthening Research Capacity and Disseminating New Findings in Nursing and Public Health – Malini et al. (Eds)*
*© 2018 Taylor & Francis Group, London, ISBN 978-1-138-50066-2*

# Challenges and strategies in conducting a health education programme for chronic disease in Indonesia

H. Malini
*Faculty of Nursing, University of Andalas, West Sumatra, Indonesia*

ABSTRACT: Providing health education to people with chronic disease such as diabetes becomes an important aspect in order to improve their quality of life. However, a structured education programme is rarely implemented because there is limited information about how such education programmes should be conducted. The aim of this paper is to discuss several aspects that have been identified during the implementation and evaluation of a proposed structured education programme (the Indonesian Group-based Diabetes Education Programme or InGDEP) in an Indonesian setting. Several aspects of the health beliefs and values of people with chronic diseases have been identified, such as those relating to traditional medicine and religious aspects. The perception of health professionals regarding the implementation of the structured education programme is also analysed. During the programme implementation, there were some challenges and strategies that have been applied and proven to be effective. It is recommended that a structured education programme should become one alternative in delivering health information to people with chronic disease in Indonesia.

## 1 INTRODUCTION

Health education is considered as an instrument of health promotion programmes that are intended to improve the health of populations and promote health (WHO, 2017). A health education programme is an activity that seeks to inform individuals of the nature and causes of illnesses such as chronic diseases. Health education is described as having several functions and actions, designed to convey health-related information, attain health-related learning, and lead to skills development and lifestyle modification (Whitehead, 2004).

A health education programme can be conducted and performed at any level and in any situation, that is, from healthy people to people with certain diseases (Glanz et al., 2008). For people with identified diseases or who have a high risk of disease, the health information is designed to help them make lifestyle choices that will lower their health risk and assist them to manage their lifestyle and disease (Glanz et al., 2008).

One of the proposed methods for delivering health education is a structured education programme. A structured health education programme can improve people's knowledge and skills and have an impact on their quality of life (Funnel et al., 2011). A structured education programme has been introduced to health professionals in *Puskesmas* (community health centres) in Indonesia with a proposed programme called the Indonesian Group-based Diabetes Education Programme (InGDEP) (Malini et al., 2017). This programme focuses on conducting regular health education sessions for a month. It is delivered by a team of health professionals and designed to suit the Indonesian cultural background and resources (Malini et al., 2017). The aim of this paper is to discuss some of the cultural beliefs, the perceptions of health professionals, and the challenges affecting the implementation of the structured education programme, as well as the strategies used to overcome obstacles within the programme.

## 2 CULTURAL BELIEFS AND VALUES OF PEOPLE WITH DIABETES

Diagnosis of diabetes disrupts the life of a person and their family. They must accept this chronic disease, learn how to adapt to the condition, and find out some necessary information in order to successfully live with it (Weinger & Leighton, 2009). How a person responds to the diagnosis depends on that person's experience, perception of diabetes, and their peer or family support, as well as the support from healthcare providers.

In terms of diabetes management, there are several cultural beliefs and values that most people with diabetes and their families hold. As in most Asian traditions, Indonesian people consider the family as an important part of their life. Individual success is most valuable in light of the honour it brings to the family. Likewise, a sickness or health condition of one family member will impact on others (Doyle et al., 2010). As in most cultures, religions provide a rich context for health beliefs and practices among Asians. Indonesians believe that illnesses are part of the destiny that the Creator (Allah) decides for them: when Allah creates some diseases, Allah also creates the cure at the same time (Wiryomartono, 2014). Traditional medicines and spiritual healings are preserved throughout the Indonesian archipelago and believed to be a part of the cure. Herbal medicines and tonics called *jamu* are both home-blended and mass-produced.

Some misperceptions regarding diabetes and its management prevail in the Indonesian community. Patients commonly assume that diabetes is caused by too much consumption of raw sugar. In their perception, when someone is diagnosed with diabetes, it is because that person consumes too much sugar. Thus, if they stop consuming sugar, the disease will be cured. Moreover, some patients think that if a diabetic has a family history of diabetes, it is natural for that person to also experience diabetes. Some people in the community also believe that diabetes can be cured after taking medication several times. A number of commercial products advertised on television, in newspapers or in pamphlets, which claim to cure diabetes, encourage such misperceptions (Candra, 2013). These misperceptions exist in the community, contributing to the increasing prevalence of diabetes, and affecting the quality of life of people with diabetes.

## 3 PERCEPTIONS OF HEALTH PROFESSIONALS TOWARDS THE STRUCTURED EDUCATION PROGRAMME

The health professionals involved in this programme have basic competence in conducting health education and they have already frequently provided educational services as educators, even though these services are not part of such a structured programme. Some factors become a consideration in the delivery process of InGDEP. After considering the workload of the health professionals, the number of patients with diabetes, and the resources and facilities of *Puskesmas*, the InGDEP is applied in the form of group-based education. To anticipate such inhibiting factors as literacy skills, language barriers, hearing impairment, and cultural barriers, all materials in the education programme were provided in audiovisual and printed forms (e.g. food pyramid model kit, booklet and leaflet, video recordings). In implementing InGDEP, the health professional team members (consisting of multidisciplinary professionals such as doctors, nurses, dieticians, and health promoters) work together with the patients and family members to enhance their knowledge and skill for diabetes self-management.

There are some perceptions that are identified from the health professionals concerning the programme. *Health professionals believed that health education is to help individuals to cope with their health condition.* The health professionals agreed that it is always possible for them to conduct a health education programme in *Puskesmas*, with the limitations and resources they have, if they can manage it well.

*Health professionals also identified that conducting the programme needs commitment and team work.* Thus, they agreed that group-based activity is the best way to conduct health education in *Puskesmas*. They also stated that support is deemed necessary from the decision makers, such as the head of *Puskesmas* and Ministry of Health officials. *Several benefits of conducting InGDEP* were identified, such as improving the knowledge and motivation of the patients, gaining new experience, and upgrading knowledge and skills in delivering health education programmes. The health education team agreed that they gained new experience, knowledge, and skills when delivering health education programmes.

## 4 CHALLENGES AND STRATEGIES IN IMPLEMENTING THE EDUCATION PROGRAMME

Some strategies needed to be implemented in order to reduce the chance of rejection of the programme. One positive point of InGDEP as a proposed educational programme is that it does not create a new programme for health professionals. The InGDEP initiative makes use of all elements that have already existed in *Puskesmas*. Basically, the implementation of the programme provides a new perspective for health professionals that educational programmes should be performed in a better way in order to have an impact on patients. Some strategies that were implemented include maintaining communication, assigning the key persons, introducing InGDEP, identifying the potential support system, and collaborating with health professionals.

The first strategy was *maintaining communications*. For this strategy, communication with the person in charge was intensively developed. The involvement of the person who is in charge of the education and the department training and is responsible for introducing new programmes from the District Health Office (DHO) to *Puskesmas* becomes the major support of the programme implementation. The second strategy was *assigning the key person* who will introduce the programme to other staff members of *Puskesmas*. Knowing the background of the team members of the educational programme was also important in understanding their point of view on certain issues. The next strategy was *identifying the potential support system* by visiting other health centres or private hospitals that conducted diabetes education programmes. The reason was to find community support for this programme's future sustainability. Identifying the potential support available for health professionals in the future is necessary if they are to maintain the educational programme. The last strategy was *collaborating with the health professionals*. In recruiting and inviting patients to get involved in this study, some strategies were implemented. The information was distributed through flyers to patients who attended *Puskesmas*. Team members with a background in health promotion also distributed the flyers and informed community members during community visits. The information in the flyers also attracted patients. The free blood check attracted patients because the test is quite expensive. To encourage the patients to complete all the education sessions, friendly reminders in the form of text messages were sent to patients a day before the session by the contact person in *Puskesmas*. In the morning, another friendly reminder was sent. This strategy was effective for most people with diabetes who did not attend the sub-district *Puskesmas* and was evidently effective in maintaining the attendance rate of the patients.

Some obvious challenges while implementing the programme concerned ways to maintain motivation. The health professionals admitted that their involvement in this project increased their motivation to do better in their job. However, maintaining their motivation over a long period to conduct the programme required some rewards and managerial support. To have financial support, the programme needs to be acknowledged by the DHO. The health professionals also mentioned that potential challenges that InGDEP would face in future would be staff member turnover and lack of resources. Staff turnover would become an issue for the team members in terms of their capability. As new staff come in, they will have different capabilities as they have not participated in the education training. Another challenge is the lack of resources, especially in terms of resources to follow up a patient's condition. Although most patients attend *Puskesmas* regularly, following up their condition would be difficult because online medical records are not available, and patients may move without any notice to other *Puskesmas*. In most cases, the health professionals would lose contact with the patients because they cannot be contacted at fixed telephone numbers or addresses.

The patients also revealed some concerns around self-management. For some participants who lived with their family members such as their children, providing the food intake suitable for them was not easy, because the participants depend on their children to prepare meals. This indicates that family member involvement in this programme needs to be enhanced, not only in accompanying the participants to *Puskesmas*, but also for preparing and measuring the food intake.

Another aspect that they felt became a barrier was in maintaining motivation and the need for support from health professionals, peers, or family to consistently perform self-care activities. Support would rarely be available if the health professionals were too busy, or peer or group meetings were not available after the participants had finished the programme. In other words, the establishment of the education programme needs to be followed up by creating some post-programme support for people with diabetes who have finished the education sessions, in order to maintain their motivation and overcome their difficulties in coping with diabetes on a daily basis.

## 5 CONCLUSION

The implementation and evaluation of an InGDEP as a structured diabetes education programme in Indonesian settings has revealed such important aspects as perception, beliefs, and values, both from the health professionals and the patients. The challenges and strategies from this programme also enrich the explanation of how the programme has been conducted. All of the aspects could become valuable information in order to improve the programme and also to anticipate several obstacles hindering the sustainability of the education programme. Improving the knowledge of people with diabetes to promote self-management behaviour can increase their quality of life. For health professionals, the InGDEP implementation provides them with valuable experience in delivering health education programmes in a structured way.

## REFERENCES


Arnold, M. (2007). Pfizer diabetes program improves health outcome. *Medical Marketing & Media*, *42*, 30–30.

Candra, A. (Ed.). (2013, June 6). *Pemahaman yang salah tentang diabetes yang berakibat fatal*. Kompas.com. Retrieved from http://health.kompas.com/read/2013/06/06/1249448/Pemahaman.Salah.Tentang.Diabetes.yang.Berakibat.Fatal.

Doyle, E.I., Ward, S.E. & Oomen-Early, J. (2010). *The process of community health education and promotion*. Long Grove, IL: Waveland Press.

Funnel MM, Brown TL, Childs BP *et al.* (2011) National standards for diabetes self-management education. *Diabetes Care*. 34: S89–95 doi: 10.2337/dc11-S089.

Glanz, K. & Schwartz, M.D. (2008). Stress, coping and health behavior. In K. Glanz, B.K. Rimer & K. Viswanath (Eds.), *Health behavior and health education: Theory, research, and practice* (4th ed., pp. 211–236). San Francisco, CA: Jossey-Bass.

Glanz, K., Rimer, B.K. & Viswanath, K. (Eds.). (2008). *Health behavior and health education: Theory, research, and practice* (4th ed.). San Francisco, CA: Jossey-Bass.

Hall, G. (2009). Successful self-management of diabetes. *Diabetes and Primary Care*, *11*(5), 286–294.

Malini, H., Copnell, B. & Moss, C. (2017). *Considerations in adopting a culturally relevant diabetes health education programme: An Indonesian example. Collegian, 24*(2), 183–190.

Skelly, A.H. (2002). Elderly patients with diabetes: What you should ask your patient on the next visit. *The American Journal of Nursing*, *102*(2), 15–16.

Weinger, K. & Leighton, A. (2009). Living with diabetes: The role of diabetes education. In K. Weinger & C.A. Carver (Eds.), *Educating your patient with diabetes* (pp. 3–14). Boston, MA: Humana Press.

Whitehead, D. (2004). Health promotion and health education: Advancing the concepts. *Journal of Advanced Nursing*, *47*(3), 311–320.

WHO. (2017, November). *Diabetes: Fact sheet*. Geneva, Switzerland: World Health Organization. Retrieved from http://www.who.int/mediacentre/factsheets/fs312/en/.

Wiryomartono B. *Perspectives on traditional settlements and communities: Home, form and culture in Indonesia.* (2014) Singapore: Springer Science & Bussiness Media Singapore.

*Strengthening Research Capacity and Disseminating New Findings in Nursing and Public Health – Malini et al. (Eds)*
*© 2018 Taylor & Francis Group, London, ISBN 978-1-138-50066-2*

# Preparing qualified nurses: Nurse education institutions' role in providing excellent standards of nurse education, graduate transition and work readiness

L. McKenna
*College of Science, Health and Engineering, School of Midwifery, Melbourne, Australia*

ABSTRACT: Quality education of nurses is vital for meeting the health care needs of society. This is a context that is constantly changing with new technologies, medicines and knowledge, requiring health providers to be adaptable and constantly updating their knowledge and skills. The context of health care requires new graduate nurses to be prepared for practice and transition smoothly into their roles into the evolving practice setting. Education institutions preparing nurses play a key role in promoting excellence in nursing care, and producing graduates who are work ready and adaptable.

## 1 INTRODUCTION

Nurse education has undergone significant transition over recent decades. We have seen a move from hospital-based apprenticeship approaches to 'training' registered nurses to tertiary-based 'education', along with the development of nursing, rather than medical-based, knowledge along with new and emerging roles for nurses. This paper explores the challenges and roles that nurse education institutions need to play to ensure excellence in nurse education, optimum graduate transition and work readiness.

## 2 GRADUATE READINESS FOR PRACTICE

Much has been written about new graduate nurse readiness for practice and the existence of a theory-practice gap between what is learnt in the education institution and that experienced in clinical practice (El Haddad et al., 2013; 2017; Wolff et al., 2010). All of that is compounded by an expectation by health care institutions for graduates to be able to 'hit the floor running', that is, to be able to assume full clinical practice roles on commencement of their employment. However, there has been a reported level of dissatisfaction with how well education providers are preparing nursing students to assume their graduate roles (Missen et al., 2016; Wolff et al., 2015). Moreover, there is anecdotal evidence to suggest that there are differences in expectations around what it means to be 'practice ready'.

## 3 CHANGING CONTEXT FOR NURSING PRACTICE AND EDUCATION

Health care information changes at a rapid pace. There are constantly new treatments, understandings of diseases, medical devices and drugs, along with increasing patient acuity (Needleman, 2013). Hence, nursing knowledge is not static and graduates need to be equipped with the skills to ensure they can respond appropriately. Furthermore, nursing can no longer be considered to be just locally specific. Global health issues, such as chronic disease, cancer, diabetes, cardiovascular disease, and HIV/AIDS, have wide ranging impact across many continents (WHO, 2017). International travel continues to grow, meaning that spread of disease is facilitated, while large numbers of refugees and migration mean that management of health issues is widespread. Such international movement not only means that nurses need to be equipped to manage various conditions, they also need to be able to provide culturally specific care to all patients. Finally, migration of nurses around the globe means that nurses educated in one setting will not necessarily practise in that place (Li et al., 2014). Hence, graduates need skills and knowledge to practise in global settings.

The nature of patients, too, has changed. No longer are patients merely recipients of health care, rather they are now informed consumers. Through the Internet, people have ready access to information and may have better understanding of their health condition than in the past (Wong et al., 2014). Patients are seeking involvement in their care through shared decision making and have expectations of their care. This, along with increasing support available on social media platforms, presents issues around accuracy and application of information to local contexts. Nurses and other health professionals need to be able to work with patients to sort through information and provide informed critique and advice (Wong et al., 2014). Furthermore, they need to accountable for their practice using evidence-based approaches to deliver person-centred care (Williams, 2010). It is no longer appropriate to do something based on the fact that it has always been done that way, rather there is a need for nurses to use evidence to support their practice, to generate evidence, and for their educational preparation to provide them with the skills to be able to achieve that. In recent years, there has been increased emphasis on patient safety and quality of care, placing further accountability responsibilities on health professionals. Much research has examined the important of teamwork and collaboration through interprofessional care in promoting quality patient care outcomes and advocacy, in order to improve morbidity and mortality. This reinforces the need for integration of interprofessional learning activities into health professional curricula (Murdoch et al., 2017).

As health care changes, so do the roles of nurses. Presently, there is increasing emphasis on leadership and advanced nursing practice roles, facilitating enhanced scope of practice. Many countries have introduced Nurse Practitioner roles and other nurse-led models of care, while many countries are seeking to increase the amount of community-based care, reducing the burden on inpatient services. The new and emerging health technologies mean that nurses need to be equipped to use and understand the technology that they employ, including electronic patient data (Needleman, 2013).

The nature of nurses themselves is changing as new generations enter the profession. The nursing workforce now comprises of Baby Boomers, Generation X, Generation Y (Millennials), and Generation Z (Post-Millennials). Generation Y are digitally connected and expect free flow of information and immediate feedback. They seek work-life balance and flexibility in the workforce (Anselmo-Witzel et al., 2017). Generation Z (Post-Millennials) are described as social media connected, technology dependent and risk averse. They prefer independent learning, more practical and hands-on (Hampton & Keys, 2017).

## 4 IMPACT ON NURSING CURRICULA

The changing nature of health care and nurses suggests the need for ongoing review and refreshing of curricula. There is a need for these to be underpinned by up-to-date research and technologies as well as responsiveness to the needs of new generations of nurses. Courses need to prepare graduates for diverse future practice, reflecting contemporary nursing and health care practice at local, national and global levels. Courses need to equip graduates to critique, use and generate research to underpin practice, and prepare graduates who can work in partnerships with health care consumers. Overwhelmingly, there is a need to provide learning opportunities for nursing students to learn alongside other health professional through incorporation of opportunities for interprofessional education.

## 5 CHALLENGES FOR EDUCATION PROVIDER

While it is clear what education providers need to do enhance graduate readiness for practice, there are many challenges to be faced. In a technology and fiscally driven health care system, we face the challenge of ensuring we retain the true 'caring' essence of what nursing is, and ensuring that consumers' and community expectations are met (Darbyshire & McKenna, 2013). School leavers today have many career choices. Nursing needs to be seen as an attractive career option with many opportunities after graduates. Often there is a perception that nursing is a career for low academic achievers and misconceptions about nursing work (Liaw et al., 2016). There is a need to work on the profile of nursing, while working to retain the existing experienced nursing workforce and continue to require excellent nurses. Maintaining flexibility within the workforce could be one way to achieve that.

To ensure our graduates are appropriately skilled, it is important to ensure that the academic nursing workforce is replenished and that academics can maintain their practice currency. Academics need to work closely with clinicians to build a supportive clinical learning context that allows effective skills development (Scully, 2011), not only in technical, but non-technical skills such as professional behaviours, caring, situation awareness (McKenna et al., 2014) and empathy. Furthermore, there is a need for ongoing clinical staff development particularly in the areas of mentoring and bedside teaching.

Academics need to build opportunities to imbed interprofessional education opportunities into the curricula of different health professions to facilitate the development of decision making and problem solving, communication and team work. In addition, there is a need to build resilience, leadership and self-management into our curricula to enhance graduates' skills in the clinical environment.

## 6 NEW AND EMERGING MODELS FOR NURSE EDUCATION

New and emerging approaches to the education of nurses offer great potential. Interprofessional education, engaging students from different disciplines in scenarios can enhance graduate preparation for a culture of working together within the health professional team. Carefully designed experiences can be strong learning motivators. In one study, McLelland et al. (2017) developed an interprofessional team-based simulation birth scenario to explore if it would improve undergraduate paramedic, nursing and midwifery students' clinical knowledge and self-efficacy scores in managing birth in an unplanned location (Emergency Department). Participants were final year undergraduate paramedic, nursing and midwifery students. Using a simulated patient, a woman in late stage labour was transported by ambulance to hospital. The birth was imminent so she was diverted to the ED. Together, the nursing and paramedic students were required to manage the woman's care until the midwife arrived. Students self-identified professional roles and scope of practice in the debriefings, while video recordings showed that students were comfortable with their own roles, but less so with others' roles. Overall, the experience was considered positive for developing better understanding of each other's roles.

Peer-assisted learning facilitates the development of teaching skills that graduates need to be able to effectively support students of nursing and other health professions in their clinical learning. A study by McKenna and French (2011) evaluated implementation of a peer-assisted learning program where final year nursing students taught first year students the skills of vital signs. Final year students developed valuable teaching skills, recognised teaching as an important part of nursing, as well as benefits for their graduate practice. First year students not only learnt the clinical skills, but experienced vicarious learning around various course aspects such as clinical placements.

It is timely also to consider new pathways for entry into nursing, such as graduate entry pre-registration programs, along with enticing offerings such as double degree courses to attract school leavers to nursing. Graduate entry masters programs in Australia have been found to attract students who may not previously have undertaken nursing courses and more males (McKenna & Vanderheide, 2012; McKenna et al., 2016). For registered nurses, there is a need to consider expanding postgraduate specialisations and research degrees: PhD, professional doctorates, honours degrees, and masters research degrees. To ensure develop student cultural awareness and understanding of global health systems, student mobility programs that entail international clinical placements (Gower et al., 2017) or study abroad opportunities through international collaborations should be made available.

## 7 CONCLUSION

In order to provide excellent standards of nurse education, graduate transition and work readiness now and into the future, nurse education academic and institutions carry a number of responsibilities. It is vital that academics work to continually refresh curriculum to meet changing health care needs and imbed new evidence. Academics, themselves, need to be research active, skilled in contemporary educational models and clinically engaged, and prepared to deliver contemporary education. They need to work to promote student inquiry and engagement. Academic institutions need to engage in meaningful interaction between health care and education sectors regarding graduate expectations and to produce adaptable, knowledgeable, questioning graduates who are equipped with extensive skills and knowledge required now and into the future for global nursing practice.

## REFERENCES


Anselmo-Witzel S, Orshan S, Heitner KL, Bachand J. 2017. Are generation Y nurses satisfied on the job? Understanding their lived experiences. *JONA: Journal of Nursing Administration* 47(4), 232–237.

Darbyshire P, McKenna L. 2013. Nursing's crisis of care: What part does nursing education own? *Nurse Education Today* 33, 305–307.

El Haddad M, Moxham L, Broadbent M. 2013. Graduate registered nurse practice readiness in the Australian context. *Collegian* 20(4), 233–288.

El Haddad M, Moxham L, Broadbent M. 2017. Graduate nurse practice readiness: A conceptual understanding of an age old debate. *Collegian* 24(4), 391–396.

Gower S, Duggan R, Dantas JAR, Boldy D. 2017. Something has shifted: Nursing students' global perspective following international clinical placements. *Journal of Advanced Nursing* 73(1), 2395–2406.

Hampton DC, Keys Y. 2017. Generation Z students: Will they change our classrooms? *Journal of Nursing Education and Practice* 7(4) 111–115.

Li H, Nie W, Li J. 2014. The benefits and caveats of international nurse migration. *International Journal of Nursing Science* 1(3), 314–317.

Liaw SY, Wu LT, Holroyd E, Wang W, Lopez V, Sim S, Chow YL. 2016. Why not nursing? Factors influencing healthcare career choice among Singaporean students. *International Nursing Review* 63(4), 530–538.

McKenna L, French J. 2011. A step ahead: Teaching undergraduate students to be peer teachers. *Nurse Education in Practice* 11(2), 141–145.

McKenna L, Missen K, Cooper S, Bogossian F, Bucknall T, Cant R. 2014. Situation awareness in undergraduate nursing students managing simulated patient deterioration. *Nurse Education Today* 34, e27–e31.

McKenna L, Vanderheide R, Brooks I. 2016. Is graduate entry education a solution to increasing numbers of men in nursing? *Nurse Education in Practice* 17(1), 74–77.

McKenna L, Vanderheide R. 2012. Graduate entry in practice in nursing: Exploring demographic characteristics of commencing students. *Australian Journal of Advanced Nursing* 29(3), 49–55.

McLelland G, Perera C, Morphet J, McKenna L, Hall H, Williams B, Cant R, Stow J. 2017. Interprofessional simulation of birth in a non-maternity setting for pre-professional students. *Nurse Education Today* 58, 25–31.

Missen K, McKenna L, Beauchamp A. 2016. Graduate Nurse Program Coordinators' perspectives on graduate nurse programs in Victoria, Australia: A descriptive qualitative approach. *Collegian* 23(2), 201–208.

Missen K, McKenna L, Beauchamp A, Larkins J. 2016. Qualified nurses' perceptions of nursing graduates' abilities vary according to specific demographic and clinical characteristics: A descriptive quantitative study. *Nurse Education Today* 45, 108–113.

Murdoch NL, Epp S, Vinek J. 2017. Teaching and learning activities to educate nursing students for interprofessional collaboration: A scoping review. *Journal of Interprofessional Care* 31(6), 744–753.

Scully, NJ. 2011. The theory-practice gap and skill acquisition: An issue for nursing education. *Collegian* 18(2), 93–98.

Needleman, J. 2013. Increasing acuity, increasing technology, and the changing demands on nurses. Nursing Economic$ 31(4), 200–202.

Williams B. 2010. The way to patient-centred care. *Nursing Management* 41(10), 10–12.

Wolff AC, Pesut B, Regan S. 2010. New graduate nurse practice readiness: Perspectives on the context shaping our understanding and expectations. *Nurse Education Today* 30(2), 187–191.

Wong C, Harrison C, Britt H, Henderson J. 2014. Patient use of the internet for health information. Australian Family Physician 43(12), 875–877.

World Health Organization. 2017. Global Health Observatory (GHO) data. http://www.who.int/gho/en/.

*Strengthening Research Capacity and Disseminating New Findings in Nursing and Public Health – Malini et al. (Eds)*
*© 2018 Taylor & Francis Group, London, ISBN 978-1-138-50066-2*

# Mental health focus: Current global public health challenges

R. Machmud
*Department of Public Health and Community Medicine, Faculty of Medicine, Andalas University, West Sumatra, Indonesia*

ABSTRACT: The Movement for Global Mental Health is a network of individuals and organisations that aims to improve services for people living with mental health problems and psychosocial disabilities worldwide, especially in low and middle income countries where effective services are often scarce. National mental health policies still do not apply a sector-wide approach. The policies should not solely concern mental disorders but should also recognise and address broader issues that promote mainstreaming mental health into policies and programmes in governmental and non-governmental sectors. Two principles (scientific evidence and human rights) are fundamental to the movement. The objective of this paper is to address the global perspective of mental health into public health responsibilities, the cultural and contextual challenges of mental health concerns in public health, and the development of a mental health policy as a public health focus.

## 1 INTRODUCTION

Mental health disorders draw a great deal of attention from countries all over the world. The gap between mental health programme achievement in developed and developing countries is becoming wider. The current health programme to address mental health problems is not satisfactory in developing countries due to the scarcity of financial and human resources, insufficient policy, plan and legislation, and low movement of the civil society (WHO, 2012; 2017). Poverty-related conditions, which characterised most of the population in low and middle income countries, were also found as significant risk factors of common mental disorders (Lund et al., 2010).

Many developing countries may face such obstacles when promoting mental health among the population. There is still a stigma in the community defining that a family with a mentally ill member will be ashamed or afraid and will then finally be forced to treat the mentally ill person through such inhuman methods as restraining them in cages, stocks, or chains (Bolton, 2014). Furthermore, governments in countries with geographic challenges of various potential disasters should also pay particular attention to a mental health programme because this aspect has appeared to be neglected (Jenkins, et al., 2011).

## 2 CHANGING THE GLOBAL PERSPECTIVE ON MENTAL HEALTH INTO PUBLIC HEALTH RESPONSIBILITIES

Mental health is a state of well-being in which an individual realises his/her own abilities, can cope with the normal stresses of life, can work productively, and is able to make contributions to his/her community (WHO, 2012). Mental health and well-being are fundamental for our collective and individual ability as humans in order to think, emote, interact with each other, earn a living, and enjoy life. On this basis, the promotion, protection, and restoration of mental health can be regarded as a vital concern of individuals, local communities, and societies throughout the world (Bolton, 2014).

The World Health Organization (WHO) has been steadily moving mental health problems up in the global ranking of causes of disability. Currently, the WHO rates depression as the single greatest cause of disability worldwide, affecting at least 350 million people. The WHO warned that depression was predicted to be the leading cause of disease globally in 2030 especially in low and middle income countries (WHO, 2017).

Mental disorders are strongly associated with an increased risk of morbidity and mortality from physical health problems, including cardiovascular disease, diabetes, HIV / AIDS, and other infectious diseases. In addition, mental disorders can have a significant and long-lasting impact on maternal and child health (Patel et al., 2004). Recent estimates even highlight the direct role of mental illness as a cause of mortality (Bolton, 2014). A growing body of evidence also points to a rapid increase of mental health problems in low and middle income countries. These mental health problems now account for the bulk of the disease burden (Kessler et al., 2009; WHO, 2012).

## 3 CULTURAL AND CONTEXTUAL CHALLENGES FOR MENTAL HEALTH CONCERN IN PUBLIC HEALTH

Mental illness carries a stigma and a negative image problem in all society. It means mental health problems remain at low priority even though they are a major cause of disability worldwide. This attitude is so widespread that one can only conclude that stigmatising those with mental disorders is a human rather than culturally defined response, although culture does appear to influence its severity and response (Bolton, 2014).

Mental health has always been treated as secondary in national health policies even in high income countries. Most countries clearly regard it with less importance than physical health. In the United States, health insurance companies have historically provided less (or no) mental health service coverage and government-supported mental health services receive a small fraction from the resources allocated to physical health services (Mental Health Parity and Addiction Equity Act 2008, 2017). In middle and low income countries like Indonesia, the subordination of mental health compared with physical health has been even more extreme. Mental health services are frequently unavailable outside cities or major hospitals. In fact, 56% of countries worldwide have no national mental health policy (WHO, 2015).

On the international level, there have been calls to increase attention to mental health and to expand the availability of mental health services in low and middle income countries (Jenkins et al., 2010; Khandelwal et al., 2010; Mbatia & Jenkins, 2010). Recent epidemiological studies indicate that globally mental disorders account for a significant proportion of public health problems (Kessler et al., 2009).

In Indonesia, especially in West Sumatra, these areas are in the western part of the Sumatra Island, which are in the earthquake-prone area which was hit by multiple earthquakes in the last decades. Regions known as disaster-prone areas may contribute to the emergence of mental health disorders (Pietrzak et al., 2012). As pointed out by many previous studies, disasters could increase the risk of mental health impairment among the population, such as through Post-Traumatic Stress Disorder (PTSD), depression, panic disorder, or generalised anxiety disorder (Pietrzak et al., 2012). WHO stated that mental health services should receive more attention from authorities after emergency services because the prevalence of mental health disorders would increase almost twofold after disasters strike as compared with the pre-disaster phase (WHO, 2012). Data from the National Survey in 2013 showed that the prevalence of mental health disorders in West Sumatra Province was higher than the national incidence. On the contrary, the proportion of the population that accessed the mental health service was low. It seems that the mental health programme has been neglected. Such high prevalence of community mental health problems will degrade the quality of the Indonesian population and will ultimately reduce the human development index.

West Sumatra's Mental Health Survey 2013 found that other causes of mental health problem in West Sumatra Province could be due to several socioeconomic factors, namely female gender, low education attainment, unemployment, and poverty. The most frequent mental health disorder among adults in West Sumatra Province was depression followed by panic disorder, anxiety, and PTSD. This finding is consistent with prior studies that stated that depression was the most common mental health disorder in developing countries particularly among females (Amstadter et al., 2009; Black et al., 2016; Patel et al., 2004). WHO also mentioned that the common mental health disorders including depression, anxiety, and PTSD, have significant contribution to the burden of diseases and disabilities in low and middle income countries (WHO, 2012). The most frequent mental health disorder in this study was also in line with the prevalence of mental health disorders in western countries (WHO, 2015). Several studies yielded evidences that major depression and anxiety disorders were among the most common mental health disorders in developing countries' population (WHO, 2015).

There are a few literatures on mental health problems. The report of West Sumatra Mental Health Survey 2013 in Indonesia will fulfil the scarcity of literature on mental health problems. This survey reported the existance prevalence of mental health disorders among children The prevalence of hyperactivity (7.2%) and autism (1.1%) reported in this study was consistent with the prevalence among children in developed countries (Black et al, 2016; Kessler et al., 2009). It also found that more than a quarter of the infant population in West Sumatra Province were at a high risk of development disorders.

These mental health disorders among children should receive high attention from the public health policy makers because one's adverse experience at a young age will lead to the high risk behaviour in one's adolescence age (Patel et al., 2004). The increasing prevalence of drug abuse, smoking, alcohol consumption, or free lifestyle nowadays may be explained by the low mental health status during childhood, which could also more probably contribute to the high occurrence of depression among adults (Black et al., 2016; Patel et al., 2004). This study's result may warn the public policy makers of the future impact of mental health disorders.

The gap for mental health programmes exists because the national mental health policies have not yet adopted a sector-wide approach. The policy should not be solely concerned with mental disorders but should also recognise and address the broader issues that promote mental health (Vreeland, 2007). These include mainstreaming mental health promotion into policies and programmes in governmental and non-governmental sectors. In addition to the health sector, it is essential to implement a sector-wide approach, such as in the education, labour, justice, transport, environment, housing, and welfare sectors (Jenkins et al., 2011).

A comprehensive action plan should recognise the essential role of mental health in achieving health for all people. It must be based on a life course approach, aim to achieve equity through universal health coverage, and stress the importance of prevention (Vreeland, 2007).

## 4 POLICY DEVELOPMENT MENTAL HEALTH AS PUBLIC HEALTH FOCUS

Mental health issues need to be seen and valued alongside physical illness. The problem of stigma and shame could become obstacles for any mental health programme, and the stigma prevalent in previous generations needs to be replaced with a culture of understanding, empathy and respect.

Mental health promotion involves actions to create living conditions and environments. It should support mental health and allow people to adopt and maintain healthy lifestyles including a range of actions to increase the chances for more people to experience better mental health (Manning, 2009).

An environment that respects and protects basic civil, political, socioeconomic, and cultural rights is fundamental to mental health promotion (Khenti et al., 2016; WHO, 2012). Without the security and freedom provided by these rights, it is very difficult to maintain a high level of mental health (Jenkins et al., 2011).

WHO supports governments in the goal of strengthening and promoting mental health. WHO is working with governments to disseminate this information and has evaluated evidence for promoting mental health to integrate effective strategies into policies and plans (WHO, 2012). To do so, the World Health Assembly approved in 2013 the so-called "Comprehensive Mental Health Action Plan for 2013–2020". The plan constitutes a commitment of all WHO's state members to take specific actions to improve mental health and contribute to the attainment of a set of global targets.

The action plan's overall goal is to promote mental well-being, prevent mental disorders, provide care, enhance recovery, promote human rights, and reduce the mortality, morbidity, and disability for persons with mental disorders. The plan focuses on four key objectives: to strengthen effective leadership and governance for mental health; to provide comprehensive, integrated, and responsive mental health and social care services in community-based settings; to implement strategies for promotion and prevention in mental health; and to strengthen information systems, evidence, and research for mental health. A particular emphasis is also given in the action plan to the protection and promotion of human rights, the strengthening and empowering of civil society, and the central place of community-based care.

Implementation of the action plan will enable persons with mental disorders to access mental health and social care services more easily and be offered treatment by appropriately skilled health workers in general health care settings. The plan will also allow them to participate in the reorganisation, delivery, and evaluation of services so that care and treatment becomes more responsive to their needs, allowing them to gain greater access to government disability benefits, housing, and livelihood programmes. Finally, it will enable those people with mental disorders to participate at work and in community life and civic affairs in a much better way.

## 5 CONCLUSIONS

Mental health is influenced by cultural and contextual challenges in public health concern. A Mental health programme should work in partnership and recognise that mental health and wellbeing is not just a health concern but a major social issue which requires a coordinated approach across all parts of the public, private and third sectors. We propose that policies to strengthen health systems should consider mental health and the health policies should include the core elements of a mental health strategy, especially the integration of mental health into primary care and the decentralisation of mental health specialist services onto the district level. This will help ensure client-focused services and strengthen primary care, which is currently facing the heavy burden of mental disorders without carefully adapted and tailored training, medicines, information systems, support, and supervision. Additionally, such policy making will also assure the expansion of access to local referrals, which are essential to achieve good outcomes for both mental disorders and their impact on physical health.

## REFERENCES

Amstadter, A.B., Acierno, R., Richardson, L.K., Kilpatrick, D.G., Gros, D.F., Gaboury, M.T. & Galea, S. (2009). Posttyphoon prevalence of posttraumatic stress disorder, major depressive disorder, panic disorder, and generalized anxiety disorder in a Vietnamese sample. *J Trauma Stress, 22*(3), 180–188. doi: 10.1002/jts.20404.

Black, R.E., Walker, N., Laxminarayan, R. & Temmerman, M. (2016). Reproductive, maternal, newborn, and child health: key messages of this volume. In R. E. Black, R. Laxminarayan, M. Temmerman & N. Walker (Eds.), *Reproductive, maternal, newborn, and child health: disease control priorities, Volume 2* (3rd ed.). Washington (DC).

Bolton, P.A. (2014). The unknown role of mental health in global development. *Yale J Biol Med, 87*(3), 241–249.

Jenkins, R., Baingana, F., Ahmad, R., McDaid, D. & Atun, R. (2011). Health system challenges and solutions to improving mental health outcomes. *Ment Health Fam Med, 8*(2), 119–127.

Jenkins, R., Kiima, D., Njenga, F., Okonji, M., Kingora, J., Kathuku, D. & Lock, S. (2010). Integration of mental health into primary care in Kenya. *World Psychiatry, 9*(2), 118–120.

Kessler, R.C., Aguilar-Gaxiola, S., Alonso, J., Chatterji, S., Lee, S., Ormel, J. & Wang, P.S. (2009). The global burden of mental disorders: an update from the WHO World Mental Health (WMH) surveys. *Epidemiol Psichiatr Soc, 18*(1), 23–33.

Khandelwal, S., Avode, G., Baingana, F., Conde, B., Cruz, M., Deva, P. & Jenkins, R. (2010). Mental and neurological health research priorities setting in developing countries. *Soc Psychiatry Psychiatr Epidemiol, 45*(4), 487–495. doi: 10.1007/s00127-009-0089-2.
Khenti, A., Freel, S., Trainor, R., Mohamoud, S., Diaz, P., Suh, E. & Sapag, J.C. (2016). Developing a holistic policy and intervention framework for global mental health. *Health Policy Plan, 31*(1), 37–45. doi: 10.1093/heapol/czv016.
Lund, C., Breen, A., Flisher, A.J., Kakuma, R., Corrigall, J., Joska, J.A. & Patel, V. (2010). Poverty and common mental disorders in low and middle income countries: A systematic review. *Soc Sci Med, 71*(3), 517–528. doi: 10.1016/j.socscimed.2010.04.027.
Manning, A.R. (2009). Bridging the gap from availability to accessibility: providing health and mental health services in schools. *J Evid Based Soc Work, 6*(1), 40–57. doi: 10.1080/15433710802633411.
Mbatia, J., & Jenkins, R. (2010). Development of a mental health policy and system in Tanzania: an integrated approach to achieve equity. *Psychiatr Serv, 61*(10), 1028–1031. doi: 10.1176/ps.2010.61.10.1028.
*Mental Health Parity and Addiction Equity Act 2008. (2017,* September 2). Retrieved from http://www.cms.gov/CCIIO/Programs-and-Initiatives/Other-Insurance-Protections/mhpaea_factsheet.html.
Patel, V., Rahman, A., Jacob, K.S. & Hughes, M. (2004). Effect of maternal mental health on infant growth in low income countries: new evidence from South Asia. *BMJ, 328*(7443), 820–823. doi: 10.1136/bmj.328.7443.820.
Pietrzak, R.H., Tracy, M., Galea, S., Kilpatrick, D.G., Ruggiero, K.J., Hamblen, J.L. & Norris, F.H. (2012). Resilience in the face of disaster: prevalence and longitudinal course of mental disorders following hurricane Ike. *PLoS One, 7*(6), e38964. doi: 10.1371/journal.pone.0038964.
Vreeland, B. (2007). Bridging the gap between mental and physical health: a multidisciplinary approach. *J Clin Psychiatry, 68 Suppl 4*, 26–33.
World Health Organisation (2012). Global burden of mental disorders and the need for a comprehensive, coordinated response from health and social sectors at the country level. Geneva: World Health Organization.
World Health Organization (2015). *Mental health atlas 2014*. Geneva: World Health Organization.
World Health Organization (2017). Deppression. *Fact sheet.* 2017 September 2, Retrieved from http://www.who.int/mediacentre/factsheets/fs369/en/.

# *Community health*

*Strengthening Research Capacity and Disseminating New Findings in Nursing and Public Health – Malini et al. (Eds)*
*© 2018 Taylor & Francis Group, London, ISBN 978-1-138-50066-2*

# Family support in relation to life quality of the elderly with hypertension

G. Sumarsih, D. Satria & D. Murni
*University of Andalas, Padang, Indonesia*

ABSTRACT: Low levels of family and social support can affect the quality of life of the elderly with hypertension. This study aimed to determine the relationship between family social support and the life quality of elders with hypertension. This was a cross-sectional study using accidental sampling technique. Data was collected from 123 elderly patients with hypertension and used the Interpersonal Support Evaluation List (ISEL)and the abbreviated World Health Organization Quality of Life (WHOQOL-BREF) questionnaires. The data was analysed using chi-squared tests. The results indicated that there is a relationship between family social support and quality of life for elderly patients with hypertension ($p$-value = 0.000). The elderly with hypertension who have high levels of family social support also have a good quality of life. Nurses are recommended to deliver counselling to families such that they should be more active in providing motivation and full attention to their elders with hypertension. As a result, it is expected that a good quality of life could be achieved.

## 1 INTRODUCTION

People who have become elderly will experience a decrease in their physical condition. This can cause disorders or abnormalities in their physical, psychological, and social functions, which in turn will cause a decrease in their quality of life. As a consequence, they will become dependent on others. Other studies have described how individuals with hypertension reported experiencing symptoms such as headache, depression, anxiety, and fatigue. These symptoms are reported to affect the quality of life of a person in various dimensions, especially the physical dimension. Therefore, in dealing with individuals with hypertension, measuring their quality of life is very important so that optimal management can be implemented.

According to the World Health Organization (WHO, 1994, cited in Mauk, 2009), quality of life is defined as an individual's perception or assessment of their position in life, in the context of cultural systems, standards, and expectations. This concept also considersphysical health, psychological state, degree of independence, social relationships, personal beliefs, as well as the environmental context.

As Azizah (2011) reported, family is the social group that has the greatest and closest emotional bond to aclient. Therefore, the family is always involved in planning, care, and treatment. Family support is a tangible subject within the social environment and affects the behaviour of the recipient. This is supported by the study in Banjar WangayaKajaof Ghita (2011), who found that 60% of family social support in the health sector was still low. This low family social support will affect the hypertensive elders' behaviour in maintaining their health and the deteriorationto their quality of life.

## 2 METHOD

This is a quantitative study that emphasises numerical data analysis. The research is designed as a descriptive study with cross-sectional approach. The cross-sectional design is a type of research that emphasises the time of measurement or observation of independent variable data and the dependent variable at only one time (Nursalam, 2013). In this study, the independent variable data consists of social support of the family and the dependent variable is the hypertensive elderly patient's quality of life. Samples in this research were obtained conveniently in that they happened to exist or were available in accordance with the research context. The total sample in this study was 123 elderly people in Andalas community health centrein 2016.

## 3 RESULTS

### 3.1 *Family social support*

The univariate analysis of the family social support variable of elderly people in Andalas-community health centre has shown that more than half (66.7%) of hypertensive elderly people receivehigh levels of social support from their family (Table 1).

### 3.2 *Quality of life*

In addition, the univariate analysis of the quality of life variable at Andalas community health centrehas similarly indicated that more than half (52%) of the elderly with hypertension have a good quality of life, as presented in Table 2.

The univariate analysis of the variable of family social support to the elderly has shown that more than half (66.7%) of hypertensive elderly people receive high levels of social support from their family. In addition, the univariate analysis of the quality of life variable similarly indicated that more than half (52%) of the elderly with hypertension have a good quality of life, as presented in Table 2.

### 3.3 *The relationship between family social support and quality of life*

To analyse the relationship between the family social support of the hypertensive elderly and their quality of life, a bivariate analysis was used to examine the relationship between independent and dependent variables. Family social support was defined as the independent variable and the quality of life was categorised as the dependent variable. A chi-squared test was applied as the bivariate test with the applied rate of error (alpha) of 5% or 0.05.

Table 3 shows that the chi-squared test resulted in a*p*-value of0.000, which indicated that there is a significant relationship between the social support of family and the quality of life of elderly hypertensive patients in Andalas *Puskesmas* (community health centre), Padang. The elderly (38 respondents or 92.7%) who had received low levels of social support from their family claimed to have a poor quality of life. Only 21 respondents (25.6%) of those

Table 1. Family social support distribution.

| Family social support | *f* | % |
|---|---|---|
| High | 82 | 66.7 |
| Low | 41 | 33.3 |
| Total | 123 | 100 |

Table 2. Distribution of elders' quality of life.

| Quality of life | *f* | % |
|---|---|---|
| Good | 64 | 52 |
| Poor | 59 | 48 |
| Total | 123 | 100 |

Table 3. Relationship between family social support and quality of life of the hypertensive elderly (n = 123).

| | Quality of life | | | | | | |
|---|---|---|---|---|---|---|---|
| Family social support | Poor | | Good | | | | |
| | *f* | % | *f* | % | Total | % | *p*-value |
| Low | 38 | 92.7 | 3 | 7.3 | 41 | 33.3 | 0.000 |
| High | 21 | 25.6 | 61 | 74.4 | 82 | 66.7 | |

claiming to have a poor quality of life had received high levels of family social support. By contrast, only three (7.3%) hypertensive old people who claimed to have a good quality of life admitted to receiving low levels of social support from their family, whereas 61 such respondents (74.4%) enjoyed high levels of family social support.

## 4 DISCUSSION

The results of our analysis showed that more than half (66.7%) of the hypertensive elderlyreceived high social support. This study echoesthe findings of Suardana et al. (2010), which showed that 45.8% of their hypertensive elderly population receiveda good level of family social support. Family social support is a process that depends upon bothfamily and social environment. This relationship covers aid in daily activities, finances, information, and coping strategies. Such family support also deals with ways to handle problems, guidance, support, recognition, appreciation and attention, and emotional support in the form of affection, trust, and listening (Friedman, 1998).

Our study also showed that more than half of the population (64 elderly hypertensive people or 52%) had a good quality of life, where as 48%(59 people) claimed to have a poor quality of life. Again, this result is in line with the study of Suardana et al. (2010), which also found 52.5% of their respondents having a good quality of life. Good quality of life means having a condition free of physical, mental, and social disease, and the ability to function in everyday life (Mubarak, 2006). In our study, we also found that less than half of these elderly respondents (38.2%) were pensioners. In this respect, Hardianti (2014) argues that there is a difference inquality of life between old people who receive pensions and those who do not. This argument is supported by our study in that 30.1% of the elderly received little money regularly and 14.6% of them did not have enough money to meet their daily needs. In addition, 68.3% of these elderly people were not satisfied with their ability to work and 61.8% of them quite often felt negative feelings such as loneliness, despair, anxiety, and depression. This result supports Azizah's (2011) argument that a person experiencing retirement will usually experience psychosocial changes such as loss of income, position, role, activity, status, and self-esteem.

Despite experiencing so many changes after retirement, elderly people are still able to play an active role and relate well with family and community (Azizah, 2011). In our study, we found that most hypertensive elders (78 respondents or 63.4%) in the sample population benefitted from the social aspect, that is, having good social relationships with family and society. Moreover, a majority of the elderly (84.6%) stated that they have good social skills.

Although the analysis reveals that more than half of the respondents had a good quality of life, low quality in some aspects can decrease the elders' quality of life, and in turn can have a negative impact on their overall life. Experiencing unsatisfactory fulfilment in some aspects of their quality of life can also generate low self-esteem, which may lead to a sense of worthlessness, inability, and lack of confidence. These psychological degradations can causethe elderly to become depressed, have suicidal tendencies, experience mental disorders, trigger self-destructive behaviour, and discourage rational ways of thinking (Hayens et al., 2008).

On the other hand, this study also discovered that a majority of the hypertensive elderly experienced a good quality of life. Such quality of life is influenced by a high level of family, social, physical, psychological, and environmental aspects. This study also supports the arguments that in order to improve their quality of life, the elderly with hypertension should be given healthy nutrition, physical activity or exercise, stress management, social and spiritual support, and cognitive or brain exercises. Physical activity can actively limit the occurrence of chronic diseases such as hypertension, which in turn may slow biological ageing, minimise the physiological effects of lifestyle, and increase life expectancy (Muhsin, 2013).

The most prominent finding in this study is the identification of hypertensive elderly (74.4%) who enjoyed optimal family social support, resulting in a good quality of life. Families provided sufficient practical, emotional, self-esteem, and information support for the elderly with hypertension such that they were able to perform daily activities normally and socialise well in their communities. Social relations are the personal relationships the elderly have with the people closest to the family. As the only and very important source in providing support, service, and comfort for many of the elderly, family members also serve as the source of support and the most meaningful help in changing the lifestyle of the elderly (Friedman, 2003, cited in Yenni, 2011). Family support is one substantial factor that can influence a person's behaviour and lifestyle, affecting their health status and quality of life. When hypertensive elderly people have a strong relationship with their family, they may be motivated to change their behaviour, which can lead to an optimally healthy lifestyle. In the end, this will improve their health and quality of life (Yenni, 2011).

This study confirms a significant relationship between social support from family and quality of life of the hypertensive elderly. There was evidently high social support from the family that affected the good quality of life of the seniors themselves and, vice versa, low levels of family support contributed to low quality of life for the elderly. Family social support is the most important social support system for the elderly people with hypertension. If we compare it with other social support systems, family social support is associated with increased self-esteem, self-confidence, life expectancy, health status, and motivation, all of which play significant roles in improving the hypertensive seniors' quality of life. Therefore, it is advisable for the family to obtain counselling through public health centres in order to be more active in providing motivation and attention to the hypertensive elderly, improving significantly the elders' quality of life accordingly.

## 5 CONCLUSION

The study has shown that more than half of the respondents at Andalas public health centre, Padang, had high family social support and experienced a good quality of life. In addition, a significant relationship ($p = 0.000$) between family social support and the elderly hypertensive patients' quality of life at Andalas public health centre was also identified.

## REFERENCES

Azizah, L. (2011). *Keperawatan lanjut usia (Elderly Nursing Care)*. Yogyakarta, Indonesia: Graha Ilmu.

Friedman, M. (1998). *Keperawatan keluarga (Family Nursing)*. Jakarta, Indonesia: EGC.

Ghita, W. (2011). *Hubungan dukungan keluarga dan perilaku pencegahan komplikasi hipertensi pada lansia* (Thesis, Department of Nursing, Denpasar Health Polytechnic, Denpasar, Indonesia).

Hardianti, H. (2014). *Pengaruh sense of humor terhadap kualitashidup pada lansia pensiunan di Kota Malang* [*The influence of sense of humour on quality of life to retired senior citizens in Malang*] (Thesis, Department of Psychology, University of Brawijaya, Malang, Indonesia).

Hayens, R.B., Frans, H.H.L. & Eddy, S. (2008). *Buku pintar menaklukkan hipertensi.* Jakarta, Indonesia: Ladang Pustaka & Intimedia.

Hipertensi. Denpasar, Bali.

Khulaifah, S., Haryanto, J. & Nihayati, H.E. (2014). Hubungan dukungan keluarga den gankemandirian lansia dalampemenuhan activitie daily livingdi Dusun Sembayat Timur, Kecamatan Manyar, Kabupaten Gresik [The correlation between family support with elderly independency in doing activity daily living]. *Indonesian Journal of Community Health Nursing*, *2*(2), 91–97.

Mauk, K.L. (2009). *Gerontological nursing: Competencies for care* (2nd ed.). Burlington, MA: Jones and Bartlett.

Mubarak, Wahit Iqbal, (2006). Buku Ajar Keperawatan Komunitas 2. Jakarta: CV. Sagung Seto

Muhsin. (2013, April 11). *Hubungan Sosialdapat Perpanjang Umur Lansia.* Hidayatullah.com. Retrieved from https://www.hidayatullah.com/iptekes/kesehatan/read/2013/04/11/2437/hubungan-sosial-dapat-perpanjang-umur-lansia.html.

Nursalam. (2013). *Konsep dan penerapanmetodologi penelitianil mukeperawatan* (3rd ed.). Jakarta, Indonesia: Salemba Medika.

Suardana WI, Saraswati NLGI & Wiratni M. (2010). Dukungan Keluarga dan Kualitas Hidup Lansia.

WHO. (1996). *World Health Organization Quality of Life.* WHOQOL. Retrieved from: http://www.who.int/healthinfo/survey/whoqol-qualityoflife/en/index2.html (Date: April, 2017).

Yenni. (2011). *Hubungan dukungan keluarga dan karekteristiklansiadengan kejadian stroke pada lansia hipertensi di wilayahkerja Puskesmas Perkotaan Bukittinggi* (Master's thesis, Faculty of Nursing, University of Indonesia, Depok, Indonesia).

*Strengthening Research Capacity and Disseminating New Findings in Nursing and Public Health – Malini et al. (Eds)*
*© 2018 Taylor & Francis Group, London, ISBN 978-1-138-50066-2*

# Comparison of health-promoting lifestyles of male and female student nurses

R.M. Sihombing & M.G.L.R. Sihotang
*Universitas Pelita Harapan, Tangerang, Banten, Indonesia*

ABSTRACT: Gender plays a role in health-promoting lifestyles amongst student nurses. Female students are more engaged in healthy behaviour and nutrition. Male students show higher levels of physical activity than females. The purpose of this study was to compare health-promoting lifestyles of male and female student nurses. This study applied a descriptive quantitative design with a cross-sectional approach. The sample included 472 nursing students. Data was collected using the Health-Promoting Lifestyle Profile II (HPLP) which consisted of 22 items translated into the Indonesian language. This study demonstrated that male student nurses had a higher score of overall HPLP than their female counterparts. Males students practised better spiritual growth and physical activity, while female students practised better health responsibility. A significant difference was found between HPLP and gender ($p = 0.003$). According to this study, it is recommended that faculty members develop interventions to assist students in improving their health-promoting lifestyle choice based on gender.

## 1 INTRODUCTION

A healthy lifestyle is related to both personal and situational factors. Age, smoking, alcohol consumption, as well as gender and ethnicity-related activities, can trigger an unhealthy lifestyle. The relationship between lifestyle choices and health outcomes has been an area of interest for many health professionals and gender differences in these choices have also been a subject of scrutiny. Males and females differ significantly in their responses to appraisals involving general state of health, hours/day engaged in social activities, frequency of drinking alcohol, amount of alcohol consumed per session, total number of sexual partners, number of meals eaten per day, participation in physical activity, completion of annual check-ups with their doctor, and screening for sexually transmitted diseases and hypertension (Dawson et al., 2007). This is similar to the results reported by Von Bothmer and Fridlund (as cited in Ying & Lindsey, 2013), where females had healthier habits relating to alcohol consumption and nutritional control, but showed no difference from males in physical activity.

Burke and McCarthy (2011) found that 20% of the students smoked, 95% consumed alcohol and 19% of the females reported that they exceeded the recommended weekly safe level for alcohol consumption. Based on a study by Hong (2007) at the University of Mahidol, Thailand, most nursing students were at a moderate level of health-promoting lifestyles. There was still a gap between the client's expectation and the nurse's real model performance. Nursing educators, as well as nursing students themselves, should raise their own concerns about health-promoting lifestyles.

Lifestyle choices have the potential to affect current as well as future health outcomes. Johannson and Sundquist (as cited in (Dawson *et al.*, 2007) reported that lifestyle factors influenced a person's health status for roughly 20 years. Results indicated that both behavioural choices and gender influenced later health outcomes. (Al-Kandari and Vidal, 2007) found that the current lifestyle choice of student nurses will ultimately affect their lives in the future. This could also potentially affect their future roles as a health promoter and role model (S, M. R. D., 2013).

Many studies have used Health-Promoting Lifestyle Profile II (HPLP) questionnaires (developed by Walker, S. N & Hill-Polerecky, D. M (1996)) in many countries. However, no study found a relationship between gender and health-promoting behaviours using a HPLP questionnaire in the Indonesian language. The purpose of this study was to determine the score of HPLP instruments and for three subscales (health responsibility, physical activity and spiritual growth).

## 2 METHOD

### 2.1 *Study design*

This study was carried out using nursing students in Tangerang, Indonesia. This was a cross-sectional descriptive study aimed at identifying the similarities and difference between male and female student nurses in their health-promoting practices.

### 2.2 *Sample/Participant/Technical sampling*

The target population was from the nursing school at Universitas Pelita Harapan, Tangerang, Indonesia with an eligibility criteria of students who had enrolled in a nursing school programme, and agreed to participate in the study. The evaluation was conducted with a total of 472 students.

### 2.3 *Ethical consideration*

Prior to data collection, ethical approval was obtained from the Ethical Committee of Mochtar Riady Institute Nanotechnology (MRIN). The informed consent of the students was obtained prior to data collection. Students were approached after their lectures and invited to participate in the study. They were advised of the purpose of the study, were told that participation was voluntary, that they had a right to withdraw at any time without an adverse impact on their study, and information remained confidential and anonymous.

### 2.4 *Data collection and analysis*

Data was collected by the use of a self-administered questionnaire at the end of the class. After completion, the students returned the questionnaire in a sealed envelope. All data was collected from 22nd to 29th August 2016. The questionnaire of the study had two sections. The first section contained questions about the demographic characteristics of the study participants, such as gender and age. Meanwhile, health-promoting behaviours of student nurses was assessed in the second section of the questionnaire. Originally, the HPLP questionnaire consisted of 52 items, with a 4-point Likert Scale from 1 (Never) to 4 (Routinely) with six theoretical dimensions of a health-promoting lifestyle. In this study, an HPLP questionnaire consisted of 22 items and three dimensions: Health Responsibility (HR, 7 items); Physical Activity (PA, 7 items) and Spiritual Growth (SG, 8 items). Every subscale was used independently from the others, while using the entire instrument gave a total health-promoting lifestyle score. A mean of $\geq 2.50$ was considered to be a positive response in line with the previous study. The questionnaire was administered to students in the Indonesian language that was adopted, with permission, from S, M. R. D. (2013). The Cronbach's alpha coefficient in this study was 0.858 for the total scale and ranged from 0.710 to 0.79 for the subscales. Data was analysed using a computer program. The Mann-Whitney test was used to compare the HPLP scores according to gender. A $p$ value of 0.05 was regarded as significant.

## 3 RESULTS

The evaluation included 472 student nurses consisting of 115 males and 357 females. Only 19 students were not included in the study because they incompletely filled out the survey. The response rate was 96.13%. The research results showed that more than three-quarters (75.6%) of nursing students were females (Table 1).

Table 1. Frequency distribution of student's gender ($N = 472$).

| Variable (gender) | Frequency | % |
|---|---|---|
| Males | 115 | 24.4 |
| Females | 357 | 75.6 |
| Total | 472 | 100 |

Table 2. Nursing student's HPLP total and subscale mean score ($N = 472$).

| | Males | | Females | | All | |
|---|---|---|---|---|---|---|
| Sub variables | Mean | *SD* | Mean | *SD* | Mean | *SD* |
| Health responsibility | 2.18 | ± 0.51 | 2.19 | ± 0.49 | 2.19 | ± 0.49 |
| Physical activity | 2.55 | ± 0.59 | 2.13 | ± 0.50 | 2.24 | ± 0.56 |
| Spiritual growth | 3.09 | ± 0.46 | 3.01 | ± 0.44 | 3.03 | ± 0.45 |
| Total HPLP | 2.62 | ± 0.41 | 2.47 | ± 0.37 | 2.51 | ± 0.38 |

Table 3. Comparison of HPLP of the respondents amongst different genders ($N = 472$).

| | HPLP | | | | | |
|---|---|---|---|---|---|---|
| | Non-healthy | | Healthy | | | |
| Variable (gender) | *n* | % | *n* | % | Total | *p* value |
| Males | 49 | 42.6 | 66 | 57.4 | 115 | 0.003 |
| Females | 212 | 59.4 | 145 | 40.6 | 357 | |
| Total | 261 | 55.3 | 357 | 44.7 | 472 | |

The results revealed that the mean item score for total HPLP of males was higher than females. The mean HPLP for the male sample was 2.62 (*SD* ± 0.40). The highest score in the subscale was 3.09 (*SD* ± 0.45) for spiritual growth and the lowest was 2.18 (*SD* ± 0.59) for health responsibility. The mean of HPLP for the female sample was 2.47 (*SD* ± 0.37). The highest score in the subscale was 3.01 (*SD* ± 0.44) for spiritual growth and the lowest was 2.13 (*SD* ± 0.50) for physical activity (Table 2).

Table 3 shows that more than half (57.4%) of the male student nurses had a healthy lifestyle however, more than half (59.4%) of female student nurses had non-healthy lifestyles. There was a significant difference between the male and female students in the overall HPLP score ($t = 0.003$, $p < 0.05$).

## 4 DISCUSSION

In this study, the male student nurses had a better HPLP than female student nurses. The male students scored more highly in their overall HPLP and in the subscales of spiritual growth and physical activity. A significant difference was found between males and females in overall HPLP. The results of this study are similar to those reported by Al-Kandari and Vidal, (2007), and Ulla Díez and Pérez-Fortis (2010). A significant difference based on gender was found. Male students displayed an overall healthier profile, exercising more frequently and manage their stress better than female students. Contrary to other studies that showed no significant differences between males and females in overall HPLP, male student nurses in this study engaged in more risky health behaviours than their female counterparts. This may be because men tend to consider themselves invulnerable to illness or injuries (Hong, no date; Lee and Loke, 2005; Dawson *et al.*, 2007; Wei et al, 2012). These contradicting results from studies may mandate the need for more research in this area.

The explanation as to why the female student nurses had better scores in health responsibility might be because females have a desire to become more knowledgeable about their own health, so many seek information that will allow them to be in control of their own health (Nies & Mc Ewen, 2015). The female students consulted doctors more frequently for their health problems when compared to male students and showed more sense of health responsibility than male students (Suraj and Singh, 2011). The students included in this study lived in a dormitory and campus that was close to the hospital. In addition, there was a freely accessible internet facility and the students were close to services that promoted health and detection of disease at an early stage. However, not many students (18%) used the internet regularly for searching for health-related articles. This implies that health is not a major agenda issue for students as far as the use of the internet or media is concerned (Suraj and Singh, 2011).

Many studies in other countries (Haddad *et al.*, 2004; Al-Kandari and Vidal, 2007; Ulla Díez and Pérez-Fortis, 2010; Wei *et al.*, 2012b) have reported that the average score of male students was higher than female students in the subscale of physical activity. In the university studied in this paper, the students have to participate in a compulsory physical education course in their first year. The university also has a sports centre which is easy to access and where the students can freely practise sporting activities. These findings suggest that continuing physical education at university may have a positive effect on health promotion and disease prevention (Wei *et al.*, 2012a).

This study found that HPLP scores for spiritual growth of student nurses in other studies, were the highest (Can *et al.*, 2008; Bryer, Cherkis and Raman, 2013; Nassar and Shaheen, 2014; Kara and Işcan, 2016). These scores are consistent with the findings of this study. Males scored higher than females in terms of spiritual growth. The influence of the culture and societal belief system might help to maintain spiritual growth, but further studies are needed to confirm this result (Al-Khawaldeh, 2014). Individuals with good spiritual development generally have a positive and optimistic outlook on their lives, are challenged to try new things, and feel a harmonious relationship with God, the universe and everything around them ( S, M. R. D., 2013 as cited in Walker & Hill-Polerecky, 1996).

There are a number of limitations to this study. Firstly, this is a descriptive prevalence study and it only consisted of students from one university. Moreover, the sample was one of convenience, which can limit the ability to generalise to other college samples. Secondly, all information was self-reported and data could have included inaccuracies due to recall bias, social desirability and sociability. Thirdly, only a small number of males participated in this study, which is reflective of the fact that nursing remains a predominantly female-orientated profession.

## 5 CONCLUSION

This study reflects that gender contributes to a health-promoting lifestyle. This study also recommended faculty members to develop interventions to assist students to improve their health-promoting lifestyle, based on gender. Additional quantitative and qualitative studies are needed to help shape a holistic picture of the health-promoting lifestyle of student nurses.

## ACKNOWLEDGEMENTS

We would like to acknowledge all of the participants in this research.

| | |
|---|---|
| Funding: | This study was supported by the Universitas Pelita Harapan. |
| Conflict of interest: | There was no conflict of interest in this study. |
| Ethical approval: | The research received ethical approval from MRIN Ethics Committee with protocol submission No. 04.1606145 on 11th August 2016. |

# REFERENCES


Al-Kandari, F. & Vidal, V.L. (2007). Correlation of the health-promoting lifestyle, enrollment level, and academic performance of college of nursing students in Kuwait. *Nursing and Health Sciences*. [Online] Vol. 9(2), 112–119. Available from: https://doi.org/10.1111/j.1442-2018.2007.00311.x.

Al-Khawaldeh, O. (2014). Health promoting lifestyles of Jordanian university students. *International Journal of Advanced Nursing Studie.* [Online] Vol. 3(1), 27–31. Available from: https://doi.org/10.14419/ijans.v3i1.1931.

Bryer, J., Cherkis, F. & Raman, J. (2013). Health-promotion behaviors of undergraduate nursing students: A survey analysis. [Online] December, 410–416. Available from: https://doi.org/10.5480/11-614.

Burke, E. & McCarthy, B. (2011). The lifestyle behaviours and exercise beliefs of undergraduate student nurses: A descriptive study. *Health Education*, [Online]. Vol. 111(3), 230–246. Available from: https://doi.org/10.1108/09654281111123501.

Can, G., Ozdilli, K., Erol, O., Unsar, S., Tulek, Z., Savaser, S., ... Durna, Z. (2008). Comparison of the health-promoting lifestyles of nursing and non-nursing students in Istanbul, Turkey. *Nursing and Health Sciences*, [Online]. Vol. 10(4), 273–280. Available from: https://doi.org/10.1111/j.1442-2018.2008.00405.x.

Dawson, K.A., Schneider, M.A., Fletcher, P.C. & Bryden, P.J. (2007). Examining gender differences in the health behaviors of Canadian university students. *The Journal of the Royal Society for the Promotion of Health*, [Online] Vol. 127(1), 38–44. Available from: https://doi.org/10.1177/1466424007073205.

Haddad, L., Kane, D., Rajacich, D., Cameron, S. & Al-Ma&aposaitah, R. (2004). A comparison of health practices of Canadian and Jordanian nursing students. *Public Health Nursing*, [Online] Vol. 21(1), 85–90. Available from: https://doi.org/10.1111/j.1525-1446.2004.21112.x.

Hong, J.F. (2007). Health-promoting lifestyles of nursing students in Mahidol University. [Online] 27–40. Retrieved from http://repository.li.mahidol.ac.th/dspace/bitstream/123456789/1500/3/ad-arboonyong-2007.pdf.

Kara, B. & Is, B. (2016). Predictors of health behaviors in Turkish female nursing students. *Asian Nursing Research Journal*, [Online] Vol. *10 (1)*, 75–82. Available from: https://doi.org/10.1016/j.anr.2015.12.001.

Lee, R.L.T. & Loke, A.J.T.Y. (2005). Health-promoting behaviors and psychosocial well-being of university students in Hong Kong. *Public Health Nursing*, [Online] Vol. 22(3), 209–220. Available from: https://doi.org/10.1111/j.0737-1209.2005.220304.x.

Nassar, O.S. & Shaheen, A.M. (2014). Health-promoting behaviours of university nursing students in Jordan. *Health* [Online] Vol. 6(6), 2756–2763. Available from: https://doi.org/10.4236/health.2014.619315.

Nies, M.A., & Mc Ewen, M. (2015) Community/public health nursing: Promoting of the health of populations, 6th edition. St. Louis, MO: Saunders/Elsevier.

S, M.R.D. (2013). A quantitative and wualitative analysis of nurses' lifestyles and community health practice in Denpasar, Bali. Thesis. Unpublished. Available from: https://digital.library.adelaide.edu.au/dspace/bitstream/2440/95135/4/01front.pdf.

Suraj, S. & Singh, A. (2011). Study of sense of coherence health promoting behaviour in north Indian students. *Indian Journal of Medical Research*, Vol. *134*(11), 645–652.

Ulla Díez, S.M. & Pérez-Fortis, A. (2010). Socio-demographic predictors of health behaviors in Mexican college students. *Health Promotion International*, [Online] Vol. 25(1), 85–93. Available from: https://doi.org/10.1093/heapro/dap047.

Walker, S. N & Hill-Polerecky, D. M (1996). Psychometric evaluation of the health-promoting lifestyle profile II. Unpublished manuscript. University of Nebraska Medical Center. Available from: https://www.unmc.edu/nursing/faculty/health-promoting-lifestyle-profile-II.html.

Wei, C.N., Harada, K., Ueda, K., Fukumoto, K., Minamoto, K. & Ueda, A. (2012). Assessment of health-promoting lifestyle profile in Japanese university students. *Environmental Health and Preventive Medicine*, [Online] 17(3), 222–227. Available from: https://doi.org/10.1007/s12199-011-0244-8.

Ying, L.I. & Lindsey, B.J. (2013). An association between college students' health promotion practices and perceived stress. *College Student Journal,* [Online] Vol. 47(3), 437–446. Available from: http://0-search.ebscohost.com.ucark.uca.edu/login.aspx?direct=true&db=a9h&AN=90516440&site=ehost-live&scope=site.

*Strengthening Research Capacity and Disseminating New Findings in Nursing and Public Health – Malini et al. (Eds)*
 *ISBN 978-1-138-50066-2*

# Factors influencing the quality of life of the Minang elderly living in nursing homes: A perspective of Minang culture

R. Sabri
*Faculty of Nursing, Andalas University, West Sumatra, Indonesia*

A.Y.S. Hamid & J. Sahar
*Faculty of Nursing, University of Indonesia, West Java, Indonesia*

Besral
*Faculty of Public Health, University of Indonesia, Indonesia*

ABSTRACT: A change in the protection of the elderly in Minangkabau has been marked by the growing number of elderly living in West Sumatran nursing homes. Nevertheless, caregivers in nursing homes have not provided the expected care as is evident from the high level of physical and psychological illness of the elderly. This study used the focused ethnographic method and aims to identify factors influencing the quality of life of the Minangkabau elderly living in nursing homes. This research involved 13 respondents who were purposefully recruited from a government-owned Panti Sosial Tresna Werdha (PSTW) in West Sumatra. This study suggests that support from PSTW management is needed to facilitate the elderly's regular meetings with their family, to arrange health checks conducted by competent parties, and provide some basic training to new caregivers in order to serve the elderly in accordance with local cultural values.

## 1 INTRODUCTION

The number of elderly in the world is growing because human life expectancy is getting longer in most countries. In 2013, the number of elderly people in the world reached 11.7% or about 841 million people and about 560 million elderly people (66.59%) lived in developing countries, including Indonesia (20.04 million or 8.05%). The number of elderly people in West Sumatra reached 8.41% of the total population. This phenomenon follows changes in the society's social paradigm to care for the elderly. The government has established nursing homes to improve the seniors' welfare but their goal has not yet been achieved.

(Setiati et al., 2011) explain that Indonesian old people's health status and quality of life can be assessed through sex, nutritional status, number of chronic diseases, functional status, and depression status. In West Sumatra, (Harni, S, 2010) also found that 38.5% of old people felt lonely in PSTW and 26.7% of the elderly residents experienced depression. The studies in West Sumatra show that elderly people have certain health problems, such as hypertension 54%, rheumatic 76%, and uric acid 24% (Austrianti, 2010). These conditions had an impact on the elderly's decreasing ability to perform their daily activities, including low self-hygiene (68%). (George et al., 2014) found that Activities of Daily Living (ADL) was one of the main aspects that had an impact on the health status and life quality of the elderly. (Sabri, 2016) also discovered that unfair attitudes when providing care, lack of communication, and action without clear explanations / arguments make the elderly feel neglected.

The attitude of caregivers who do not appreciate the elderly is contrary to Minangkabau cultural value that upholds and respects parents, especially elderly women. Caregivers may fail to comply with the principle of *adat basandi syarak, syarak basandi kitabullah*

(Minangkabau traditional rules are based on Sharia and Sharia rules are based on Al-Qur'an) indicating that there must be no deed or action from any member of Minangkabau society who are characterised as *masyarakat sakato* (a united society) that should violate Minangkabau traditional and Islamic rules. Such a typical characteristic of Minangkabau society becomes the key assumption the researcher used in discovering causes for the caregivers' failure to adequately communicate with the elderly residents in PSTW. This lack of communication discourages the elderly to complain and tell the caregivers what they wanted (Zwygart-stauf-facher, 2007). Unfulfilled hopes and unsatisfied needs may lead the elderly into depression and this is a high risk condition that impacts their quality of life. This research identifies factors influencing the quality of life of the Minangkabau elderly people living in nursing homes.

## 2 METHOD

This study applies the ethnographic approach. Participants involved in this research were recruited through the purposive sampling technique. Data was obtained from 13 participants through observation, Focus Group Discussion (FGD), and interviews that were conducted from May to July 2016. Participants were asked some questions regarding the reason for staying at home or moving to nursing home, the changes felt before and during their stay in the nursing home, the social support they expected, and the elderly person's opinion of the behaviour of nurses and caregivers. Data analysis was conducted manually using the following steps: 1) encoding data for descriptive Table; 2) sorting data to recognise patterns; 3) identifying extreme data; 4) explaining patterns and data by using theory and theoretical construction; 5) making memos or notes to be included in the reflective statement. Results of the analysis are focused on answering the research questions (Roper & Shapira, 2000). They are presented in the form of themes that are discussed based on ethic and emic perspectives.

## 3 RESULT

### 3.1 *Theme 1: Family and peer group support*

Support of the family is really needed by elders who live in nursing homes. However, from five elders who were interviewed, only one elder claimed to have been visited by his family members. More than half of the Minangkabau elderly in the nursing homes have children but their children were migrating or there were conflicts that had happened between them so that the elderly decided to live in the nursing homes. However, during the observation, the researcher never saw any family visiting their elders. This phenomenon is supported by the caregivers' explanation:

> …He left his children because he has to go to work in Jakarta. He did not contact his family for so long and in fact he had already got married.
> …His children feel so angry and they do not want to take care of Mr. O.
> (A caregiver).

This explanation is supported by one of the elders that was interviewed by researcher:

> …I don't want to bother my family, especially my kids. It is because they usually become angry at me if I inhibited their job. I thought that it was better for me to stay here.

Almost all families of the elderly did not fulfil their parent's needs. The families said that it was because they did not have any money to buy what the elderly relative needed. Having inadequate support from the family, the elders also admitted to losing their pride as indicated by Mr. W:

> ...Living in this nursing home is like we are thrown away from the family. My pride is falling down. I felt like falling from the 40th floor, far from the children, families, and my life. I cannot have any interaction with the people here.

Support does not only come from family but also from fellow elders in the same nursing home. This support can take positive or negative forms. The research found that such positive and negative supports mostly came at the same time, as described by Mr.U:

> ...friend is always there until they reach what they wanted. In addition, I and my friends are having different backgrounds. ...sometimes they said bad things about me in front of the caregivers so the caregivers got angry at me.

Sometimes, conflicts happened between the elders in the nursing home and their discussion topics were always about those conflicts. Mr. Z added to this situation:

> ...we are just like kids here. Sometimes, it is more than the kids' behaviour. Sometimes, they told the caregivers but the caregivers never responded to it and just smiled. That is why we just can stay silent and we will not inform the caregivers again because it has no impact.

### 3.2 *Theme 2: Life expectancy*

For an elder living in the nursing home, his expectations do not match reality. There are many things that make them complain about living in nursing homes, such as objections to some programmes, never considering the ideas and opinions of the elderly, and not being addressed by nurses and caregiver. Furthermore, they are only told to do their activities. Fatigue and boredom can lead to depression in the elderly, which ultimately affects the quality of their lives. The doubt is reflected through their statement:

> We just accept any instructions. We are not allowed to give any opinions or idea. We are commonly having problems here.... We accept what they gave us. We may not be asking too much.
>
> (Mr. SI).

A similar statement, Mr. SI who said that, that at first, he thought that living in a nursing home would make him feel better, but the reality did not match his expectations:

> ...but I never found any activity that makes me comfortable. It seems that my hopes are not reached yet again. If we talked about hopes, I did not have any hopes anymore.
> (Mr. U, Mr. Sm and Ms. I).

Hope does not match reality. That is a fact found in a PSTW owned by the local government. Nurses and caregivers have behaviors that do not match the expectations of the elderly. The fear of asking caregivers has worsened their condition, degrading their life expectancy, as Mr. J and Mr. S. stated:

> "...not all the staffs meet my hopes..."... there is a staff who always uses a loud voice when calling me. Sometimes, when the food is not appropriate and I told them, they asked me to buy the food by myself.
>
> (Mr. J).

### 3.3 *Theme 3: Caregivers' attitude*

The caregivers' attitudes also affect the quality of life of the elderly. The results of the observation show that there is only one caregiver who is trusted by the nursing home residents. The other caregivers did not give careful attention and enthusiastic services to the elderly. This negative attitude was confirmed by the senior's confessions such as Mr. D:

> ...I am already old. I need a suitable medicine from my respiratory illness. When I complained about this, they always asked me to sleep, take a rest, and not to walk too much. I already did those things. I told them that I needed medicine but they said that I had to buy the medicine by myself.

Mr. K, who had lived for about three years in the nursing home, also gives the same statement:

...My illness is getting worse because these caregivers were not caring for me wholeheartedly. I wanted to go to the mosque for praying together, listen to some lectures, or read Qur'an together. However, those caregivers did not help me to go to the mosque.

Because of some physical changes, their inabilities for carrying out daily activities adds to the caregivers' job list. These unwanted tasks affect the caregivers' attitudes to the elderly, which causes pressure for the elderly because they cannot get what they want.

## 4 DISCUSSION

Family members are the closest persons to elderly people because they are the only persons who take a responsibility to accompany the elderly in their old age. Family members and peer groups are two of the informal social support for the elderly besides the formal support obtained from the nursing home (Gottlieb & Bergen, 2010; Mattson & Hall, 2011). However, the relationship between the elderly with families and their friends can take negative or positive forms of support. Negative support was still found among elders living in nursing homes. (Azwan et al., 2015) explain that the support from friends can have different impacts on the life quality of the elderly living in the nursing home, whether their relationships are negative or positive. From the perspective of Minangkabau culture, friends of the same age are partners for the elders to share their family and life experiences. The elderly will feel that getting together with friends can be like when they were with their family (Sabri, 2016).

Minangkabau culture had defined the appropriate attitudes on how to be in a society and to communicate with people from all ages, including with friends of the same age. Different understandings of the terms *"kato nan ampek"* (four ways of having words with others) covering *kato mandaki* (ways of having words with older people), *kato malereng* (ways of having words with a respected person), *kato manurun* (ways of having words with younger people) and *kato mandata* (ways of having words with same-aged people) (Amir, 2011), sometimes can trigger conflicts between one elderly and the others, between the elders and caregivers. These conflicts create pressure for the senior residents of the nursing home, which can result in uncomfortable feelings (Chang, 2013). Chen et al., (2007) argued that a good family relationship with caregivers in nursing homes will improve the life quality of the elderly because families will know the quality of services given by caregivers in caring for their parents.

Commonly, Minangkabau elderly people live in nursing homes located close to their houses but the ignored old people are usually taken to government social institutions. Although families of the elderly usually come from unhealthy families that cannot support their old people financially, they must give emotional support, reward, information, and social integration. Optimum support from the family and friends will improve the elderly's hope in the nursing home. Some elderly people feel comfortable living in the nursing homes, because they have no family and permanent residence.

Ghimire and Gurung (2014) explain that the elderly's life quality in nursing homes is also influenced by their families' visits The value of continuous family visits in nursing homes, the role, and support from the family are really important for the elderly's quality of life. (Chen et al., 2007) propose that seniors who live in nursing homes do have hopes for love, trusted feeling, great services, and friendly services as much as when they live with their family. Amir (2011) also said that friendliness and love are the Minangkabau cultural forms of *sahino samalu* (equally ashamed and embarrassed) and *raso jo pareso* (having awareness and self-correctness).

Health conditions also have a certain impact on the quality of life of the elderly. Lack of special attention in terms of food services becomes a source of complaints that affect the health status of the elderly. Farzianpour, et al., (2015) explain that elders with low physical conditions will have low abilities for performing their daily routines. This influences their emotional and social relationships with caregivers. The cultural Minangkabau perspective of Minangkabau cultural values *barek samo dipikua* (heavyweight thing is carried together on shoulders), *ringan samo dijinjiang* (lightweight thing is carried equally by hand) means that the people always do good things, help each other, give advice, and respect togetherness. Therefore, the elderly people who have limitations will not feel useless because they are mutually *sehino semalu* (have the same position in the nursing home).

By doing so, caregivers would have to implement the Minangkabau rules of *Nan buto pahambuih lasuang* (the blind will blow and keep the camp fire up); *nan pakak palapeh badia* (the deaf will release the canon); *nan lumpuah pahuni rumah* (the crippled will stay and watch for the house); *nan bingung ka disuruah-suruah* (the confused will be guided to help); *nan pandai tampek batanyo* (the intelligent will become the source of information); *nan cadiak bakeh baiyo* (the smart will become a discussion partner) and *nan kayo tampek batenggang* (the rich will accommodate complaints and provide tolerance). These principles define proportional job distributions in accordance with the individuals' expertise and ability (Amir, 2011; Naim, 2013).

## 5 CONCLUSION

The position of the elderly in Minangkabau is as *pai tampek batanyo* (someone to ask when we want to go), *pulang tampek babarito* (someone to inform when we want to return home). This research has identified three main themes related to changes in Minangkabau cultural values that degrade the quality of life of elderly people living in nursing homes. Implementation of Minangkabau cultural values in services for the elderly, especially in nursing homes, needs to be addressed and reconceptualised. The implications of this study for nurses and caregivers in providing services to the elderly are able to increasing knowledge and skills in providing care to the elderly in nursing homes, so that the elderly feel comfortable and protected. Nursing and caregiver nurses need to be aware of the elderly's wishes and expectations while living in a nursing home.

## REFERENCES


Amir, M.S. (2011). The minangkabau tradition: pattern and purpose of people's lives minang 7th edn. Edited by M. Pabayuang. Jakarta: Citra Harta Prima.

Austrianti, R. (2010). The relationship between depression levels with ability levels to conduct an activity daily living at Panti Sosial Tresna Werdha (nursing Home).. Padang. Retrieved from: http://www.repository.unand.ac.id/18336.

Azwan, Herlina, Karim, D. (2015). The relationship between social support of peer group with elderly quality of life at Panti Sosial Tresna Werdha (nursing Home). *Journal Online Mahasiswa*, *2*(2), 962–970.

Chang, S.J. (2013). Lived experiences of nursing home residents in Korea. *Asian Nursing Research*. Elsevier, *7*(2), 83–90. doi: 10.1016/j.anr.2013.04.003.

Chen, C.K., Sabir M, Zimmerman S., Suitor J., & Pillemer K. (2007). The importance of family relationships with nursing facility staff for family caregiver burden and depression., *The Journal of Gerontology*, *62*(5), 253–260.

Farzianpour, F., Foroushani, A. R., Badakhshan, A., Gholipour, M., & Roknabadi E. H. (2015). Quality of life for elderly residents in nursing homes. *Global Journal of health Science*, *8*(4), 127–135. doi: 10.5539/gjhs.v8n4p127.

George, P. P., Heng, B. H., Wong, L.Y., & Ng, C.W. (2014). Determinants of health-related quality of life among community dwelling elderly. *Annals Academy of Medicine Singapore*, *43*(1), 3–10.

Ghimire, S. & Gurung, S. (2014). Role of family in elderly care. (Unpublished Master Thesis). School of Health Care and Social Services Degree programme in nursing. Lapland University of Applied Scienties.

Gottlieb, B.H. & Bergen, A.E. (2010). Social support concepts and measures. *Journal of Psychosomatic Research*, 69(5), 511–520. doi: 10.1016/j.jpsychores.2009.10.001.

Harni, S.Y. (2010). The relationship psychosocial factors with depression in elderly at panti sosial tresna werdha (nursing homes) Sabai Nan Aluih Sicincin Padang Pariam*an*. Padang. Retrieved from http://repository.unand.ac.id.

Mattson, M. & Hall, J.G. (2011). *Health as communication nexus: A service learning approach. 1*st edn. Kendall Hunt Publishing Co. Retrieved from https://he.kendallhunt.com/product/health-communication-nexus-service-learning-approach.

Naim, M. (2013). *The wander: The migration pattern of Minangkabau ethnic.* 3rd ed. Jakarta: Rajawali Press. Retrieved from http://www.rajagrafindo.co.id.
Roper, J.M. and Shapira, J. (2000). *Ethnography in nursing research*: What to do with all the data. ISBN: 0761908749. Thousand Oaks-London-New Delhi: Sage Publications Inc.
Sabri, R. (2016). An overview of quality of life of minangkabau elderly, in the living at nursing homes. In Proceedings of the National Seminar on Nursing and Scientific Presentation. Palembang: IKA-PERSI dan BEM PSIK Unsri, 78–86.
Setiati, S. *et al.* (2011). Predictors and scoring system for health-related quality of life in an Indonesian community-dwelling elderly population. *Acta medica Indonesiana*, 43(4), 237–42. Retrieved from http://www.ncbi.nlm.nih.gov/pubmed/22156355.
Setiawan, A. et al. (2016). The statistics of the elderly population year 2015. In S. Susilo, D., Harahap, & I., E., Yasmuarto (Eds.). Jakarta: Badan Pusat Statistik. Retrieved from https://bps.go.id/website/pdf_publikasi/Statistik-Penduduk-Lanjut-Usia-2015—.pdf.
Zwygart-stauffacher, M. (2007). 'Measuring quality of care in assisted living', *Journal of Nursing Care Quality, 22*(I), 4–7.

*Health management*

*Strengthening Research Capacity and Disseminating New Findings in Nursing and Public Health – Malini et al. (Eds)*
*© 2018 Taylor & Francis Group, London, ISBN 978-1-138-50066-2*

# The supervision experience of head nurses in a hospital setting

D.S. Paramitha
*University of Muhammadiyah Banjarmasin, Banjarmasin, South Borneo Province, Indonesia*

ABSTRACT: Supervision performed by a head nurse is an activity that provides direction and guidance to improve nursing care quality. In Indonesia, there is still little information that reveals in depth how the head nurse conducts supervision. This study aims to explore various experiences of head nurses in implementing supervision by using a qualitative method with a phenomenological approach at Ratu Zalecha Hospital. The data were collected by in-depth interviews with six head nurses as participants, selected by a purposive sampling technique. Four themes were identified by thematic analysis, namely, the perception of self-performance as a head nurse, the process of supervision, obstacles faced in implementing supervision, and the expectations desired of supervision activities. Based on the findings, the head nurses were aware of the capabilities required in supervision, and also the importance of increased support from directors of nursing for the application of effective supervision.

*Keywords*: experience, head nurse, supervision

## 1 INTRODUCTION

The head nurse, as a first-line manager, should be able to perform nursing management functions to ensure the quality of nursing care. One nursing management function is the directing function (Marquis & Huston, 2015). In the meantime, supervision is one important part that cannot be ignored in the directing function. Research suggests that effective supervision involves providing professional accountability, skill and knowledge development, and also colleague/social support (Brunero & Stein-Parbury, 2008). Supervision is one dimension of improving the quality of nursing skills (Cruz, 2011).

This phenomenon is the background of this research to explain more deeply how the practice of supervision is implemented by the head nurse. Subjective perceptions and diverse experiences of head nurses are difficult to quantify (Afiyanti & Rachmawati, 2014). Until now in Indonesia, especially in hospitals of the Banjar District, South Kalimantan Province, there has been little qualitative research that reveals the experience of head nurses from their personal accounts. Based on the above description, the purpose of this study was to obtain an overview of supervision practices conducted by head nurses.

## 2 METHOD

### 2.1 *Study design*

This research uses a qualitative method with a phenomenological approach, as it was expected that various behaviours based on reality and the views of head nurses about supervision could be described. This research was conducted from November 2016 to January 2017 at Ratu Zalecha Hospital, Martapura, South Kalimantan Province, Indonesia.

### 2.2 *Participant selection*

There were six participants recruited by a purposive sampling technique. They were head nurses who were still active and had experience as head nurses in the wards for at least three years. Participants consisted of three men and three women. The age range of participants was 42–49 years. The length of service as head nurse ranged from 5–16 years. The educational background of the participants was quite varied, with two Registered Nurses, one Bachelor of Nursing, two Bachelors of Public Health, and one Bachelor of Applied Science with a midwifery background.

### 2.3 *Ethical consideration*

Approval for this research was received from the Ethics Committee of the University of Muhammadiyah Banjarmasin, approval number 291/UMB/KE/XI/2016. Ethical considerations in this research are based on four basic principles of research ethics, namely, respect of the person, beneficence, non-maleficence and justice. The researcher respect the autonomy of participants to decide will or will not participate in, protect the identities and keep the data confidential. The informed consent sheet was signed by the participants after the researcher explained the research objectives, research procedures, duration of participant's involvement, and participant's rights.

### 2.4 *Data collection and analyses*

Data were obtained through in-depth interviews. Interviews were conducted with semi-structured techniques and in the form of open questions. The guide questions are '*How do you see yourself as the head nurse?*' and '*What do you do in the supervision activities?*' Records of interviews were then organised in the form of verbatim transcripts. Verbatim transcripts were shown to participants to ensure the comparability between the transcript data and what the participants have said. Furthermore, as this research was analysed using thematic analysis, verbatim transcripts were formed into themes to reveal the meaning of each participant's statement (Creswell, 2013). The point of saturation of the research was felt when the responses from one participant were the same as the other participants and repetition of the answers occurred.

## 3 RESULTS AND DISCUSSION

### 3.1 *The perception of self-performance as a head nurse*

This first theme is built from the participants' views of themselves, that as head nurse they must be aware of their capabilities. Participants considered that they should have the ability to perform nursing actions properly in order to be able to direct other nurses to work according to standards. This was seen from the following participant expression:

> *I have to know all the skills around the procedures; yes, there is a standard operating procedure.* (Head Nurse 5)

Furthermore, five participants agreed that the basic difference between being an associate nurse and head nurse is the number of responsibilities that must be carried out, such as expressed by a 44-year-old male participant:

> *Well, first in terms of responsibility. We have more responsibility.* (Head Nurse 1)

Ultimately, their abilities and responsibilities make them capable of becoming role models, as revealed by the following participant with a midwifery background:

> *Yes, because I am expected to be a role model, and set a good example.* (Head Nurse 2)

Based on the findings, head nurses realise the capabilities they must have. This is in line with the results of research that indicated that supervisors need enough clinical and administrative knowledge and skills for their roles, which should be considered in their training and preparation (Dehghani et al., 2016). It is also in accordance with research that says supervision is a guide to personal development, professionalism and knowledge (Kilminster et al., 2007). Therefore, the head nurse is a role model who is not only responsible for nursing skills but also ensures that the professional values of the nurse are embedded in the nurse's behaviour. The head nurse becomes an example of this behaviour in the course of the task to support the development of nursing skills (Brunetto et al., 2011).

### 3.2 *The process of supervision*

The second theme is the process of supervision, which consists of supervision planning, supervision implementation, and monitoring and evaluation of supervision. In the supervision planning phase, participants state that they make a supervision schedule. In supervision planning, the head nurse also determines what topics will be evaluated. Usually the topic is about unskillful execution of nursing procedure by nurses. This was highlighted by the following participant with a Bachelor of Nursing background:

> *I usually divided the supervision. There's a schedule. I make a schedule for supervision of the team leader; for example, today for team leader 1, tomorrow for team leader 2.* (Head Nurse 1)

The existence of a supervision planning phase shows that prior to starting supervision, the head nurse will decide in advance what supervision will be done, who will be supervised, how to do it, when it will be done, and where the supervision activities are done. Research reveals that good planning functions will improve the quality of nursing care (Safitri et al., 2013). Therefore, it is impractical for the head nurse to plan something that cannot be achieved or is abstract.

In the supervision implementation phase, the participants claimed that they try to supervise in stages. This was expressed by a female participant with a Registered Nurse background:

> *I usually ask the team leader to supervise those under him/her.* (Head Nurse 6)

In the supervision implementation phase, all participants teach skills and knowledge development to their nurses. In addition, participants also revealed that they emphasise professional attitudes and show a sense of care as a form of social support in work. This is revealed by the youngest participant, with a Bachelor of Public Health background:

> *We try to do one action by practising directly, for example, we use a syringe pump so in one meeting we all learn together ... Ethically, every day we give briefings, advice and accompany them in giving a service.* (Head Nurse 3)

The last phase is monitoring and evaluation of supervision. In this phase, the participants described their evaluation of the skill, discipline and job done by the nursens. This was expressed by a female participant:

> *If it began to decrease again, I ask the patient, there are some nurses who still do not know or understand. So, if there is a decrease, I will evaluate it again.* (Head Nurse 6)

Supervision activities conducted by the head nurse show that they teach skills and knowledge development. Based on these findings it can be assumed that the head nurse who is involved directly will observe and provide feedback to the associate nurse. The findings show that there is a relationship between the head nurse and associate nurse. This relationship, at the interpersonal level, affects the quality of the supervision process (Sivan et al., 2011). This demonstrated that they are sharing their time and other resources such as information, knowledge and skills, support and assistance with one another (Brunetto et al., 2013). Therefore, it can be assumed that the head nurse shows mutual support, empathy and understanding of the character of the staff, so as to provide comfort in working. This finding is in line with research which states that, in the monitoring and evaluation phase, the head nurse controls and identifies the constraints/problems that arise (Rashed et al., 2015). Evaluation is intended to improve responsibility and a nurse's commitment to work.

### 3.3 *Obstacles faced in implementing supervision*

The third theme relates to the obstacles faced in implementing supervision. Workload is identified as the main factor that inhibits the implementation of supervision. In addition, the lack of knowledge about management has led to the emergence of self-imposed barriers when it comes to carrying out supervision. Take, for example, the experience of the oldest participant, who showed a lack of confidence due to an educational background that was only a Nursing Diploma but continued his studies to the bachelor's degree in public health sciences. The participant said:

> *The rules of the game are that the supervisor must have a nursing education at least to Registered Nurse. I have a lot of shortcomings. I'm not suitable because my educational background is just a diploma.* (Head Nurse 4)

This head nurse feels that a lack of knowledge about management results in supervision not being optimally implemented. Additionally, there are participants who do not feel confident because their educational background is not in accordance with the requirement to be a head nurse. This is in agreement with research that states that the head nurse who is confident in sharing knowledge and experience with their staff will make those staff feel welcome in their environment (Mather et al., 2015). The number of meetings or workload is also used as an excuse not to supervise. Research suggests that the constraints are the difficulty of finding someone who is capable of supervision, time problems, high workload, and lack of preparation (de Abreu & Marrow, 2012). Therefore, the implementation of supervision requires competence as a head nurse, knowledge of supervision, a sense of responsibility, and sufficient time.

Meanwhile, another obstacle faced in the implementation process of supervision is that there is no optimal support from top nursing managers. A female participant even smiled when explaining her opinion that so far there has been no support from the nursing managers in the hospital:

> *Smile first [participant smiles] ... There is no support [laugh] ... The managers want us to do things, but so far, we just do so by ourselves.* (Head Nurse 5).

### 3.4 *Desired expectations of supervision activities*

The fourth theme is the desired expectations associated with supervision activities. The expectations desired by participants in supervision activities certainly include the ability to change things for the better. Another expectation is that of increased cooperation, increased knowledge, support for the provision of facilities and infrastructure, as well as appropriate rewards. However, the most important expectation of participants is that top nursing managers control and give clear direction about the nursing service process in the ward, as expressed by one participant:

> *We hope there is a little reward, yes. If it's not possible every month, maybe at least every three months they come here and direct us about nursing ....* (Head Nurse 5)

Based on the findings, to build a culture of supervision also requires support from top nursing managers, such as participation in staged supervision activities, provision of guidance for supervision, improving knowledge about supervision practices, the availability of facilities and support for supervision, as well as appropriate rewards. Support, in the form of rewards, both materially and morally, is actually a component of employee satisfaction. Dissatisfaction can cause a sense of disappointment in the absence of response and support, a situation that will contribute to declining performance. Research shows that there is a relationship between perceptions of organisational support and employee engagement. A good relationship between a head nurse and an associate nurse will increase their commitment to work (Brunetto et al., 2011).

## 4 CONCLUSIONS

Supervision activities involve planning, implementation, monitoring and evaluation. Experiences are influenced by perceptions of self-performance as a head nurse, obstacles, and support from top nursing managers. This research can be used as a foundation in identifying the needs for the knowledge development of head nurses regarding supervision, the need for continuing education for nurses, the availability of support facilities and infrastructure, and the support from top nursing managers to be directly involved in supervision activities. Future research could involve exploring the implementation of supervision at several hospitals, both public and private, so that the findings can be generalised.

## REFERENCES


Afiyanti, Y. & Rachmawati, I.N. (2014). *Qualitative Inquiry Methods in Nursing Research*. Jakarta, Indonesia: Rajawali Pers.

Brunero, S. & Stein-Parbury, J. (2008). The effectiveness of clinical supervision in nursing: An evidence based literature review. *Australian Journal of Advanced Nursing, 25*(3), 86–94.

Brunetto, Y., Farr-Wharton, R. & Shacklock, K. (2011). Supervisor-nurse relationships, teamwork, role ambiguity and well-being: Public versus private sector nurses. *Asia Pacific Journal of Human Resources, 49*, 143–164.

Brunetto, Y., Shriberg, A., Farr-Wharton, R., Shacklock, K., Newman, S. & Dienger, J. (2013). The importance of supervisor-nurse relationships, teamwork, wellbeing, affective commitment and retention of North American nurses. *Journal of Nursing Management, 21*(6), 827–837.

Creswell, J.W. (2013). *Qualitative Inquiry & Research Design: Choosing Among Five Approaches.* Thousand Oaks: Sage Publication Ltd.

Cruz, S.S.S.M.S. (2011). Clinical supervision in nursing: Effective pathway to quality. *Procedia—Social and Behavioral Sciences, 29*, 286–291.

de Abreu, W.J.C. & Marrow, C.E. (2012). Clinical supervision in nursing practice: A comparative study in Portugal and the United Kingdom. *Sanare, 11*(2), 16–24.

Dehghani, K., Nasiriani, K. & Salimi, T. (2016). Requirements for nurse supervisor training: A qualitative content analysis. *Iranian Journal of Nursing and Midwifery Research, 21*(1), 63–70.

Kilminster, S., Cottrell, D., Grant, J. & Jolly, B. (2007). Effective educational and clinical supervision. *Medical Teacher, 29*(27), 2–19.

Marquis, B.L. & Huston, C.J. (2015). *Leadership roles and management functions in nursing: Theory and application* (8th ed). Philadelphia, PA: Wolters Kluwer.

Mather, C.A., McKay, A. & Allen, P. (2015). Clinical supervisors' perspectives on delivering work integrated learning: A survey study. *Nurse Education Today, 35*(4), 625–631.

Rashed, S.A.E., Al Torky, M.A.M. & Morsey, S.M. (2015). Performance of head nurses management functions and its effect on nurses' productivity at Assiut University Hospital. *IOSR Journal of Nursing and Health Science (IOSR-JNHS), 4*(5), 38–49.

Safitri, N., Widiyanto, P. & Marta, R. (2013). The relationship of head nurses management functions with quality of intravenous infusion insertion procedure in Muntilan Hospital. *Proceedings of National Seminar and Call For Paper Independence in Creating and Innovating Health.* Magelang, Indonesia: University of Muhammadiyah Magelang.

Sivan, M., McKimm, J. & Held, S. (2011). Can an understanding of transactional analysis improve postgraduate clinical supervision? *British Journal of Hospital Medicine, 72*(1), 44–48.

*Strengthening Research Capacity and Disseminating New Findings in Nursing and Public Health – Malini et al. (Eds)*
*© 2018 Taylor & Francis Group, London, ISBN 978-1-138-50066-2*

# Development and validation of a Technological Competency as Caring in Nursing Instrument (TCCNI): Learning from the outcomes

I. Yuliati
*Catholic School of Health Sciences, St. Vincentius a Paulo, Surabaya, Indonesia*

J.D. Lorica
*St. Paul University Philippines (SPUP), Tuguegarao City, Cagayan, Philippines*

ABSTRACT: Caring in nursing practice is the essence of the human health experience. Caring is embedded in the human relationship between the carer and the cared for. However, there is a wide gap between the technology and human caring experiences. The use of technology often disregards the human sense of taking proper care for the cared for. A combination of technologies and caring in nursing is a harmonious aspect of nursing care. The purpose of this study is to carry out a systematic review of the existing literature from 2011–2017 and identify the development and validation of Technological Competency as Caring in Nursing Instrument (TCCNI). A multiple database search was conducted and 12 articles were reviewed. This study shows the development and validation of a nursing instrument process. The technological competency is an expression of caring. The study was able to frame the existing literature that facilitated in the explanation of the development and validation of an instrument.

## 1 INTRODUCTION

Measurements become part of nursing practice in order to make decisions. Furthermore, research in every discipline is encouraged to develop instruments or develop guidelines to measure complex concepts or constructs. Good instruments will help the researcher gain information in accordance with their wishes or expectations. Therefore, the perfect method and process of instrument development is needed.

Both nursing practice and nursing education are facing the implementation of technology. However, the perception of using technology and caring varies widely. Technologies and caring in nursing can be a harmonious aspect in the field of nursing but technology cannot replace the care received from nurses.

Therefore, knowing that technological competency in nursing is an expression of caring (Locsin, 2016) should be understood by nurses. The aim of this study was to conduct a systematic review of the existing literature from 2011–2017 to identify the development and validation of a Technological Competency as Caring in Nursing Instrument (TCCNI). This literature review can provide more information about the "translation, adaptation, and validation of an instrument or scales for use in cross-cultural health care research", that is, TCCNI.

## 2 METHOD

The Preferred Reporting Items for Systematic Review and Meta-Analyses (PRISMA) guidelines were used for reporting on the systematic literature review. Using the keywords "development and validated technological competency as caring in nursing", 237 scholarly journals were found in ProQuest and 13 articles found from the EBSCO Cumulative Index

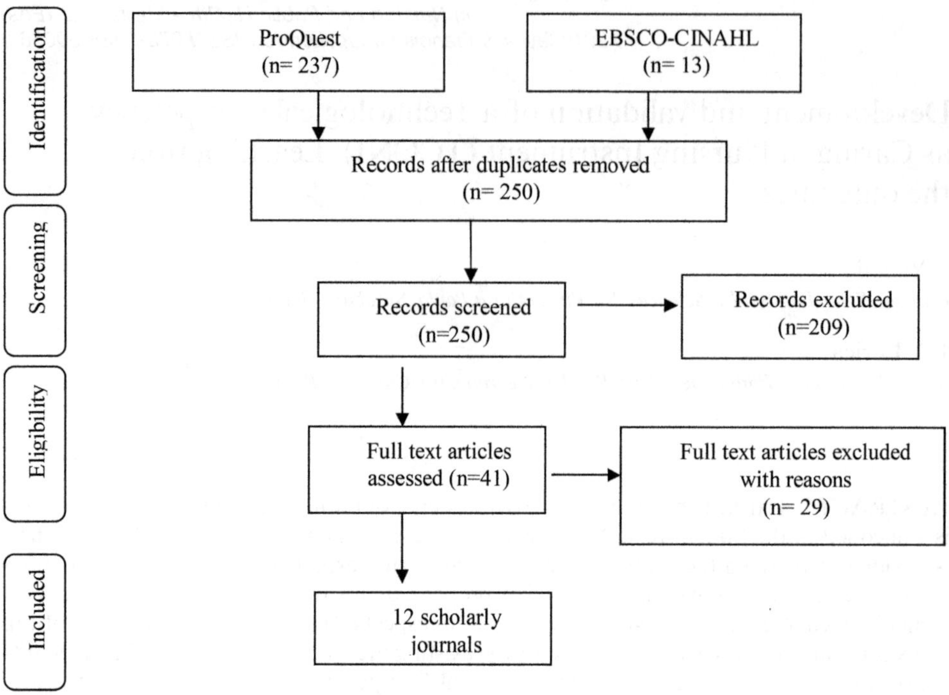


Figure 1. Diagram of review process based on PRISMA.

to Nursing and Allied Health Literature (EBSCO-CINAHL) database. The inclusion criteria were: (a) publicationin English; (b) international journals or publishing journals; (c) focused on the process of development and validation of an instrument in nursing. The result was 12 articles to be reviewed. These articles were summarised according to their content.

## 3 RESULTS & DISCUSSION

### 3.1 *Technology and caring*

Technology plays an important role in healthcare and helps nurses provide comprehensive nursing care to patients. There was a positive association between caring attributes and the influences of technology among nurses. Female nurses had a better attitude towardsthe influences of technology on their care. In contrast, younger and less experienced nurses had negative views on the effects of technology on nursing care (Bagherian et al., 2017). Technology supports the lives of critically ill patients and brings experienced nurses very close to their patients/families.

Technology allows nurses to know and assist the patient more fully in their well-being. However, technology, such as computers and other healthcare technologies, have the potential to impede the nurses' decisions and abilities to confirm patient' health condition.

Technology provides advance information and communication, which are significant factors influencing healthcare systems. Technology is sometimes understood in the context of cause and effect relationships. However, to adopt only the binary framework can diminish and obscure the allure of technological efficiency in creating human caring realities. Technology has become integral to human well-being, and the journey beyond the bedside, beyond the community, and even beyond the planet; therefore, technology and caring must coexist in nursing. Technological competency as an expression of caring in nursing is real and opportunistic, much like the mythical phoenix arising from the ashes. Technology plays a major role in caring for critically ill patients. Its use in patient care can improve patient safety, save patients' lives, facilitate nurses' work and save them time, and reduce hospital costs.

### 3.2 *Development of an instrument process*

There are several steps in, and approaches to, the development of an instrument, such as assessing content validity (considered one of the most critical steps). A developmental stage includes domain identification, item generation, and instrument formation. A judgement-quantification stage occurs when experts review the items and either report validity of the items subjectively or with an empirically referenced method, such as calculation of the content validity index. According to Parcells and Locsin (2011), the development process of an instrument is through construct quantification. The Delphi method is used to establish the content validity of an assessment instrument (Viviers et al., 2016). A dual-method approach, modified-Delphi technique using an expert clinician panel and semi-structured one-on-one interviews with a purposive sample are required for the development of the most relevant process of the care checklist items (Conroy et al., 2013). Another study explained the methods used for the development of an instrument are descriptive and cross-sectional research design and, furthermore, the four-step approach(extensive literature reviews, thematic analysis of literature, expert opinion in the questionnaire items, and psychometric properties) indicated a valid and reliable instrument (Al Hadid et al., 2011). A mixed-method approach with individual, structured cognitive interviews and quantitative assessment are used to validate the tool (Skúladóttir & Svavarsdóttir, 2016).

### 3.3 *Learning from the outcomes*

This literature review shows several results or findings regarding caring, technology and the process of instrument development. Technological competency is needed for nurses to be able to provide care for their patients. There was a positive association between caring attributes and influences of technology among nurses (Bagherian et al., 2017). We also found there were beneficial effects of technology on nursing care, which are saving patients' lives, saving nurses' time, and improving decision-making.

## 4 CONCLUSION

Caring is the essence, moral idea, and foundation of professional nursing (Boykin & Schoenhofer, 2001). Technology is not merely about machines or devices, it is about saving human lives.

This systematic review was able to frame the existing literature that facilitated the explanation of the development and validation of an instrument. From this study of the literature,the researchers used the Delphi method most often. The Technological Competency as Caring in Nursing Instrument (TCCNI) was last validated in 2011. It has been revised and 25 items have been adopted for online use with a visual analogue scale. This TCCNI has not been applied in Indonesia. Therefore, translation, cross-culturaladaptation and validation of instruments or scales in the Indonesian setting is recommended.

## REFERENCES

Al Hadid, L.A., Hasheesh, M.A.& Al Momani, M. (2011). Validating a tool that explores factors influencing the adoption of principles of evidence-based practice. *Journal of Nursing Education*, *50*(12), 681–687.

Bagherian, B., Mirzaei, T., Sabzevari, S.&Ravari, A. (2016). Caring within a web of paradoxes: The critical care nurses' experiences of beneficial and harmful effects of technology on nursing care. *British Journal of Medicine and Medical Research, 15*(9), 1–14.

Bagherian, B., Sabzevari, S., Mirzaei, T.&Ravari, A. (2017).Effects of technology on nursing care and caring attributes of a sample of Iranian critical care nurses. *Intensive and Critical Care Nursing, 39*, 18–27.
Beck,C.T.& Gable,R.K. (2001).Ensuring content validity: An illustration of the process. *Journal of NursingMeasurement,* 9(2),201–215.
Boykin A. & Schoenhofer S.O. (2001) Nursing as a caring: A model for transforming practice. Jones & Bartlett, Sudbury.
Conroy, K.M., Elliott, D.& Burrell, A.R. (2013). Developing content for a process-of-care checklist for use in intensive care units: A dual-method approach to establishing construct validity. *BMC Health Services Research, 13*(1), 380.
Dorigan, G.H.&Guirardello, E.B. (2013). Translation and cross-cultural adaptation of the Newcastle satisfaction with nursing scales into the Brazilian culture. *Revista da Escola de Enfermagem da USP, 47*(3), 562–568.
Im, E.O.& Chang, S.J. (2012).A systematic integrated literature review of systematic integrated literature reviews in nursing. *Journal of Nursing Education, 51*(11), 632–640.
Kikuchi, J. (2004). The binary: An obstacle to scholarly nursing discourse? *Nursing Philosophy, 7*, 100–103.
Kongsuwan, W. & Locsin, R.C. (2011). Thai nurses' experience of caring for persons with life-sustaining technologies in intensive care settings: A phenomenological study. *Intensive and Critical Care Nursing, 27*(2), 102–110.
Locsin, R.C. (1995a). Technology and caring in nursing. In A. Boykin (Ed.), *Power, politics and public policy: Amatter of caring*(pp. 24).New York, NY: National League for Nursing Press.
Locsin, R.C. (1995b). Machine technologies and caring in nursing. *Journal of Nursing Scholarship, 27*(3), 201–203.
Locsin, R.C.(2016). Technological competency as caring in nursing: Co-creating moments in nursing occurring within the universal technological domain. *Journal of Theory Construction andTesting,20*(1), 5–11.
McGrath, M.(2008). The challenges of caring in a technological environment: Critical care nurses' experiences. *Journal of Clinical Nursing, 17*(8), 1096–1104.
Nguyen, D.N., Zierler, B.& Nguyen, H.Q. (2011). A survey of nursing faculty needs for training in use of new technologies for education and practice. *Journal of Nursing Education, 50*(4), 181–189.
Páez-Esteban, A.N., Caballero-Díaz, L.P., López-Barbosa, N., Oróstegui Arenas, M., Orozco-Vargas, L.C. & Valencia-Angel, L.I. (2014). Construct validity of an instrument to assess patient adherence to antihypertensive treatment. *Enfermería Global, 1*(34), 48–57.
Parcells, D.A. & Locsin, R.C. (2011). Development and psychometric testing of the technological competency as caring in nursing instrument. *International Journal for Human Caring, 15*(4), 8–13.
Queirós, S.M.M., de Brito Santos, C.S.V., de Brito, M.A.C. & Pinto, I.E.S. (2015). Development of a form to assess the self-care competence of the person with a tracheostomy. *Revista de Enfermagem Referência, 4*(7), 51–59.
Ruzafa-Martinez, M., Lopez-Iborra, L., Moreno-Casbas, T.& Madrigal-Torres, M. (2013). Development and validation of the competence in evidence-based practice questionnaire (EBP-COQ) among nursing students. *BMC Medical Education, 13*(1), 13–19.
Skúladóttir, H. & Svavarsdóttir, M.H. (2016). Development and validation of a clinical assessment tool for nursing education (CAT-NE). *Nurse Education in Practice, 20*, 31–38.
Stevens, K. & Palfreyman, S. (2012). The use of qualitative methods in developing the descriptive systems of preference-based measures of health-related quality of life for use in economic evaluation. *Value in Health, 15*(8), 991–998.
Viviers, M., Kritzinger, A. & Vinck, B. (2016). Development of a clinical feeding assessment scale for very young infants in South Africa. *South African Journal of Communication Disorders, 63*(1), 1–11.
White, K.A. (2013). Development and validation of a tool to measure self-confidence and anxiety in nursing students during clinical decision making. *Journal of Nursing Education, 53*(1), 14–22.

*Strengthening Research Capacity and Disseminating New Findings in Nursing and Public Health – Malini et al. (Eds)*
*© 2018 Taylor & Francis Group, London, ISBN 978-1-138-50066-2*

# A phenomenological study: The first clinical experience of student nurses in Padang

Nelwati
*Faculty of Nursing, University of Andalas, Padang, West Sumatra, Indonesia*

ABSTRACT: This study was conducted to explore, understand and give meaning to the first clinical experience of student nurses undertaking clinical education in Padang. As a qualitative piece of research, this phenomenology study was conducted on six participants who met the criterion, were selected by a purposive sampling technique, and participated in an in-depth interview. The interviews were performed in a convenient setting for the participants and recorded. The interview was then transcribed verbatim and thematic analysis was used to examine the responses. Four themes were identified: 'being fearful'; 'having a gap between theory and practice'; 'being proud to help patients'; and 'feeling abandoned'. It is suggested that nursing educational institutions pay attention to the preparation programme for initial clinical education, so that the learning process can run optimally.

## 1 INTRODUCTION

Indonesian nursing education programmes at undergraduate level consist of two stages: academic and professional. The former is conducted in classrooms and laboratories while the latter takes place within clinical settings. Clinical education is compulsory for student nurses during their nursing learning programme. It is conducted to achieve a set of competencies, integrate the theory with practice, enhance critical thinking and decision-making abilities, and prepare the students to become professional nurses (Chapman & Orb, 2001; Chesser-Smyth, 2005; DeYoung, 2009). As a consequence, clinical education should be a worthy experience in order to achieve the expected competencies (Nelwati et al., 2013).

Clinical education is defined as 'the provision of teaching in health sciences in the form of lectures, demonstration, individual instruction, etc., and the supervision and assessment of practical application of therapeutic and patient care techniques' (Penguin Books & Macquarie Library, 1989, p. 63). In addition, Emerson (2007) stated that clinical education had to equip student nurses with comprehensive learning experiences that reflected nursing practice. Therefore, nursing teaching staff should manage the clinical education of student nurses properly.

Clinical learning processes and experiences take place differently. Neill et al. (1998) assessed the clinical experience of American student nurses during a clinical placement. A focus group discussion and observation was conducted to collect data from 75 sophomore student nurses. The findings suggested that student nurses who had no experience seemed doubtful in their roles and competencies. Thus, observation was one of their learning process methods. Those student nurses absolutely depended on their mentors and sought information from each other because they needed reinforcement (Neill et al., 1998).

Positive clinical experience during clinical education affects student nurses' perceptions on the nursing profession. Peyrovi et al. (2005, p. 137) conducted a study to examine and understand the clinical education of Iranian student nurses. Five student nurses who had finished their clinical practice participated in this study. Five themes emerged from the study, which included 'caring-oriented relationships, attractive aspects of clinical experience, finding oneself in [the] clinical milieu, being supported by classmates, and actualising potential'.

Peyrovi et al. (2005) suggested that Iranian student nurses' experiences in clinical education were interesting and useful because they contributed to the application of theory in clinical settings. Positive experiences supported them in developing caring attitudes and becoming professional nurses. On the other hand, Pearcey and Draper (2008) found that the experiences of student nurses in England was far removed from their expectations. This was caused by too many written assignments, routine work in the hospital, and difficulty in building relationships; these student nurses failed to develop a caring attitude.

During their clinical education, student nurses can also face negative experiences that lead to anxiety and stress (Gorostidi et al., 2007; Mahat, 1998; Sheu et al., 2002; Shipton, 2002). Literature consistently shows that student nurses do experience stress in their clinical education, especially during their first clinical experience. According to Admi (1997), the nature of their first clinical experience is determined by previous experience, but age was not related to this first experience. Furthermore, Sheu et al. (2002) found that stress in the first clinical experience had both a positive and negative impact on the biopsychosocial status of student nurses in Taiwan.

The first clinical experience can be different among student nurses. Chesser-Smyth (2005) explored the initial clinical experience of student nurses in Ireland. An in-depth interview was conducted to collect data from nine female students and one male student. Data were analysed using the Colaizzi approach and the five main themes that emerged were 'self-awareness, confidence, anxiety, facilitation, and professional issues' (Chesser-Smyth, 2005, p. 320). Beck (1993, p. 489) also explored the first clinical experience of student nurses in the United States and found six main themes, namely, 'pervading anxiety, feeling abandoned, encountering reality shock, envisioning self as incompetent, doubting choices, and uplifting consequences'.

Meisenhelder's study (as cited in Chesser-Smyth, 2005) found that if the first clinical experience of student nurses produced anxiety and stress, it would hinder the learning process due to a narrowed perception. As a result, student nurses should experience their first clinical encounter optimally. Thus, it is necessary for educational institutions to understand the meaning and essence of the first clinical experience for student nurses during their clinical education. Accordingly, the purpose of this study was to examine, explore and understand the meaning of the first clinical experience of student nurses in Padang, Indonesia.

## 2 METHODS

A qualitative research type and a phenomenological approach was chosen for this study to explore people's experience in real-life and to then give them meaning (Houser, 2008; Polit & Beck, 2004; Speziale & Carpenter, 2007). The phenomenon studied was the first clinical experience of student nurses during clinical education in Padang.

The qualitative research technique used depends on in-depth information taken from the selected student nursing participants who have experienced the phenomenon. To this end, a purposive sampling technique was applied. To determine the number of participants, the principle of data saturation was used (Polit & Beck, 2004; Speziale & Carpenter, 2007). As a result, six student nurses (P1–P6) participated in this study.

Prior to a semi-structured interview, potential participants were provided with an explanatory statement of the study and informed that their participation was absolutely voluntary, free from coercion, and would have no impact on their study. If they agreed to take part, a written informed consent stating that they agreed to be interviewed, and that it would be audio-taped, was subsequently signed. Confidentiality and anonymity of data obtained were also assured. Furthermore, pseudonyms are used in the reporting of data in this study in order to protect individual identities (P1–P6).

In-depth and semi-structured interviews lasting from 15 to 20 minutes were conducted in order to explore the first clinical experience of the student nurses. The question 'Could you tell me what was your experience during your first clinical education?' was used. Interviews were recorded with an MP4 player and conducted in a comfortable place. During interview sessions, participants' gestures and body language was also noted.

Data analysis took place simultaneously alongside data collection; a recorded interview was listened to repeatedly and then transcribed verbatim. The transcription was read several times (Holloway & Wheeler, 2002). Thematic analysis was used to identify the main themes and thus give meaning to the experiences.

To ensure the trustworthiness and credibility of this study, interviews were transcribed by the researcher and then sent back to the participants to check the accuracy of the transcriptions. No participant requested any change be made.

## 3 RESULTS AND DISCUSSION

Four main themes emerged from this study: 'being fearful; 'having a gap between theory and practice'; 'being proud to help patients'; and 'feeling abandoned'. The 'being fearful' response was caused by fear of incorrectly performing a clinical procedure on patients during their hospital care. Because of this initial experience, student nurses felt unconfident to perform clinical procedures. As one participant said:

> *During the first clinical education, we were not allowed to perform clinical procedures. So, when we were asked to do a particular procedure, we were afraid to make contact with clients. Even to take vital signs, we were in doubt.* (P4)

Another participant admitted:

> *I was asked to remove the infusion set by a ward nurse. However, as I had not yet learned about how to do this, I was anxious and fearful.* (P2)

A further participant confessed:

> *When I performed the first injection ... I was afraid to make a mistake and hurt the client. It was also frightening to be responded to negatively by clients.* (P6)

Fear was also experienced by almost all student nurses during their initial clinical experience. Being worried that they would make mistakes, and feeling unconfident in performing their nursing skills, gave rise to feelings of fearfulness. This finding is congruent with the research findings of Mahat (1998) and Sheu et al. (2002). It also supports the findings of Nelwati et al. (2013) in which student nurses felt fearful that they would make mistakes, hurt patients, and lack the necessary competence. Thus, student nurses experienced stress and felt pressurised during their clinical education (Nelwati et al., 2013). Being fearful in the clinical settings initiated anxiety and this hindered their learning capacity (Locken & Norberg, 2005). Consequently, fearful and anxious experiences during their first clinical education should be minimised so that the learning process is not disrupted. Furthermore, student nurses may act as a passive observer at first, but then they become an active learner, and are ready to obtain the competencies in their clinical placement (Chesser-Smyth, 2005).

All participants stated that they were 'experiencing a gap between theory and practice' because they found a gap between the theory they had learned in the classroom and the laboratory, with what they faced during their initial clinical education. As a participant said:

> *It was so different ... what we learned at campus was different from what we saw in the clinical area. For example, when we performed bed making, we made a triangle at the corner of the blanket but a ward nurse shouted at us that it should be just rolled and tied.* (P1)

Another participant informed the researcher that:

> *At campus, we learned such clinical procedures as theory, but at the clinical setting, we were just asked to take patients to the X-ray room and pick up the oxygen box. So, what did we learn the theories for?* (P4)

The other participants explained:

*In the field, theory was not used … it was just neglected. In every clinical procedure, we put on gloves for our safety but a ward nurse said that we did not need to wear them when performing procedures on the clients.* (P5)

A gap between theory and practice was found frequently by student nurses during clinical education both at academic and practice stages in their learning process (Sharif & Masoumi, 2005; Syahreni & Waluyanti, 2007). This also appeared during the first clinical education of student nurses in Padang. According to Syahreni and Waluyanti (2007), clinical education is the integration between theory and practice which enables the students to apply their acknowledge attained from the academic stage. However, a gap between theory and practice was also found due to the limitation of nursing theory implementation, and adjustment to the situation and condition in Indonesia (Syahreni & Waluyanti, 2007).

Five out of six participants gave meaning to their initial clinical experience as 'being proud to help patients'. Student nurses were proud to help patients when their patients trusted them and assumed them to be a 'real nurse'. A participant illustrated this with:

*There was a patient calling me "Sir" during clinical education and it made me feel like a real nurse.* (P2)

Another participant said:

*This should be you in the future … becoming a nurse who is called to come to assist the patients.* (P3)

However, one participant was rejected by the patient because of her status as a 'student' nurse. The patient did not trust her and refused care from her. This could make the student nurses feel ashamed and incompetent, as a participant described:

*There was also a time when I was not accepted by patients who asked "Are you a student nurse?" At that time, I took the blood pressure of the patient and told him the result and he was not convinced with the result and asked me to call a real nurse.* (P5)

Even though the participants felt afraid of making mistakes when providing nursing care, they felt proud to help patients who considered the students as real nurses. This is congruent with the findings of Syahreni and Waluyanti (2007) that student nurses who were recognised during their clinical education as real nurses were satisfied and proud of helping patients. The recognition given to student nurses as a health team member in clinical settings could enhance the clinical learning process. In contrast, when a student's attendance in a clinical setting was not fully accepted, it was found that the achievement of clinical competencies could be hindered (Syahreni & Waluyanti, 2007).

The participants gave the meaning of 'feeling abandoned' to their first clinical education because their academic clinical instructor arrived late on the first day; this upset the students. They felt that the academic clinical instructor did not guide and teach them properly. The attendance of an academic instructor is compulsory from a student nurses' view point. Without a guidance process during clinical education, especially on the first day, the student nurses felt abandoned. Therefore, it is necessary to establish good communication and a relationship between student nurses and academic clinical instructors. Harmonious relationships with their instructors could improve the clinical learning process (Ferguson, 1996; Syahreni & Waluyanti, 2007). The initial clinical learning process also requires intensive guidance because, by referring to Emerson (2007), the model for the learning process during initial clinical education is the traditional model where the instructors facilitate the process by accompanying the students to achieve a set of competencies.

## 4 CONCLUSION

An analysis of the experiences of student nurses during their first clinical education showed that they felt fearful and abandoned, encountered a theory–practice gap, and were proud to be recognised as real nurses. Therefore, it is necessary for educational nursing institutions to prepare and pay attention to new arrangements in the initial clinical education stage. As a consequence, the students would undertake the learning process more positively and be more confident in the clinical setting.

## ACKNOWLEDGEMENTS

The author would like to thank all the participants who shared their valuable experiences for this study. This research was funded by the Faculty of Nursing, University of Andalas, Padang, West Sumatra, Indonesia.

## REFERENCES

Admi, H. (1997). Nursing students' stress during the initial clinical experience. *Journal of Nursing Education, 36*(7), 323–327.

Beck, C.T. (1993). Nursing students' initial clinical experience: A phenomenological study. *International Journal of Nursing Studies, 30*(6), 489–497.

Chapman, R. & Orb, A. (2001). Coping strategies in clinical practice: The nursing students' lived experience. *Contemporary Nurse, 11*(1), 95–102.

Chesser-Smyth, P.A. (2005). The lived experiences of general student nurses on their first clinical placement: A phenomenological study. *Nurse Education in Practice, 5*(6), 320–327.

DeYoung, S. (2009). *Teaching strategies for nurse educators* (2nd ed.). Upper Saddle River, NJ: Pearson Education.

Emerson, R.J. (2007). *Nursing education in the clinical setting*. St. Louis, MO: Mosby Elsevier.

Ferguson, dS. (1996). The lived experience of clinical educators. *Journal of Advanced Nursing, 23*(4), 835–841.

Gorostidi, X.Z., Egilegor, X.H., Erice, M.J.A., Iturriotz, M.J.U., Garate, I.E., Lasa, M.B. & Cascante, X.S. (2007). Stress sources in nursing practice. Evolution during nursing training. *Nurse Education Today, 27*(7), 777–787.

Holloway, I. & Wheeler, S. (2002). *Qualitative research in nursing* (2nd ed.). Oxford, UK: Blackwell Science.

Houser, J. (2008). Nursing research: Reading, using and creating evidence. Sudbury: Jones and Bartlett Publisher:

Locken, T. & Norberg, H. (2005). Reduced anxiety improves learning capacity of nursing students through utilization of mentoring triad. *Journal of Nursing Education, 48*(1), 17–23

Mahat, G. (1998). Stress and coping: Junior baccalaureate nursing students in clinical settings. *Nursing Forum, 33*(1), 11–18.

Neill, K.M., McCoy, A.K. & Parry, C.B. (1998). The clinical experience of novice nursing students in nursing. *Nurse Educator, 23*(4), 16–21.

Nelwati, McKenna, L. & Plummer, V. (2013). Indonesian student nurses' perceptions of stress in clinical learning: A phenomenological study. *Journal of Nursing Education and Practice, 3*(5), 56–65.

Pearcey, P. & Draper, P. (2008). Exploring clinical nursing experiences: Listening to student nurses. *Nurse Education Today, 28*, 595–601.

Penguin Books & Macquarie Library. (1989). *Penguin Macquarie dictionary of health sciences*. Ringwood, Australia: Penguin.

Peyrovi, H., Yadavar-Nikravesh, M., Oskouie, S.F. & Bertero, C. (2005). Iranian student nurses' experiences of clinical placement. *International Nursing Review, 52*(2), 134–141.

Polit, D.F. & Beck, C.T. (2004). *Nursing research: Principles and methods* (7th ed.). Philadelphia, PA: Lippincott Williams & Wilkins.

Sharif, F. & Masoumi, S. (2005). A qualitative study of nursing students experiences of clinical practice. *BMC Nursing, 4*(6), 1–7.

Sheu, S., Lin, H.S. & Hwang, S.L. (2002). Perceived stress and physio-psycho-social status of nursing students during their initial period of clinical practice: The effect of coping behaviors. *International Journal of Nursing Studies, 39*(2), 165–175.

Shipton, S.P. (2002). The process of seeking stress-care: Coping as experienced by senior baccalaureate nursing students in response to appraised clinical stress. *Journal of Nursing Education, 41*(6), 243–255.

Speziale, H.J.S. & Carpenter, D.R. (2007). *Qualitative research in nursing: Advancing the humanistic imperative* (4th ed.). Philadelphia, PA: Lippincott Williams & Wilkins.

Syahreni, E. & Waluyanti, F.T. (2007). Nursing students experiences during clinical practice. *Jurnal Keperawatan Indonesia, 11*(2), 47–53.

*Strengthening Research Capacity and Disseminating New Findings in Nursing and Public Health – Malini et al. (Eds)*
*© 2018 Taylor & Francis Group, London, ISBN 978-1-138-50066-2*

# Nursing students' coping mechanisms against bullying during nursing professional practice

Z.M. Putri, I. Erwina & Y. Efendi
*Faculty of Nursing, Andalas University, Padang, West Sumatra, Indonesia*

ABSTRACT: Coping mechanisms determine a person's response to a stressor, either short-term or long-term stressors. Nursing students often experience some stressors while doing their clinical practice, such as lack of knowledge and clinical experience, and have problems managing good relationships with clinical counsellors or those who arein contact with the students during clinical practice. This study is intended to determine the correlation between bullying and nursing students' coping mechanisms. This is a correlational analytic research with a cross sectional approach. The results showed that there was a correlation between bullying and the coping mechanism in nursing practice ($p = 0.000$). It was expected that the students were able to make efforts to prevent the bullying incident by improving their knowledge and skills while practising as well as improving their socialisation and effective communication skills in dealing with others.

## 1 INTRODUCTION

Nursing education is one division of health education whose orientation is to arrange an effort to improve the quality of the professional nurse through education (Nursalam & Ferry, 2008). While undergoing the educational process, nursing students acquire knowledge and skills through theoretical and practical learning activities. Clinical practice is the most essential element in nursing education. However, when performing nursing practice, students cannot be sheltered from various problems or stressors.

Some stressors that students often experience when performing clinical practice include lack of knowledge and clinical experience, concern at making mistakes, lack of self-esteem, unfamiliarity with the clinical environment, and, most importantly, obstacles in maintaining good relationships with clinical counsellors or people who are in contact with the students while completing clinical practice. One aspect of these practice-related problems is the occurrence of negative behaviours such as bullying (Khater, Laila Zaheya & Insaf Shaban, 2014; Rahman, 2014).

According to Strauss (2012), bullying is a negative behaviour that is carried out by one or more people repeatedly overtime, which is aimed at a person who has difficulties in defending him/herself and is characterised by an imbalance of power. In nursing education, bullying is defined as repetitive behaviour which occurs in health and education facilities and is performed by one or more persons against others in order to intimidate, humiliate, and offend the victims (Smith et al., 2015).

Bullying in the nursing student context can fall under several forms of action. In general, bullying is classified into three types; verbal, physical, and psychological (Sejiwa, 2008; Werner, 2012; Chakrawati, 2015).Studies about bullying ofnursing students have been undertaken in several countries. Research conducted by Rahman (2014) at Damanhour University in Egypt discovered that 88% of 772 nursing students had experienced bullying behaviour. This data is also supported by research by Clarke et al. (2012) where 88.72% of674 nursing students in Canada encountered a bullying action.

Bullying that is experienced by the students will result in numerous negative consequences. It produces not only physical but also psychological effects on the student's performance in the form of insomnia, fatigue, declining physical health, anger, uneasiness, anxiety, stress, hatred, helplessness, lack of trust, the urge to leave work or the profession, concentration dissipation, motivation waning, and social relation disorder (Seibel, 2013; Rahman, 2014).

The response to bullying will involve a variety of coping mechanisms. Research conducted by Cooper et al. (2011) discovered that, from 1,133 students in 20 nursing schools in the southern states of the United States, around 73.4% used a coping mechanism with passive behaviour, 43.5% used active behaviour, and 18.1% indicated aggressive behaviour.

A coping mechanism is any effort that is directed to the management of stress, including the effort to solve urgent problems and the defence mechanisms to protect oneself (Muhith, 2015). Coping mechanisms areestablished through learning and reminiscing processes. Learning means the ability to adapt oneself (adaptation) toward the influence of internal and external factors (Nursalam & Ninuk, 2011).

The use of coping mechanisms depends on how individuals deal with problems. There are two types of coping mechanisms which are commonly used by individuals. The first one is called an adaptive coping mechanism, which is defined as the coping mechanism which supports the function of integration, growth, learning, and achieving the goals. The second one is called a maladaptive coping mechanism, which constitutes the coping mechanism that restrains the function of integration, breaks growth, decreases autonomy, and tends to dominate the environment (Stuart, 2013).

Clarke (2009) argues that nursing students tend to use the adaptive coping mechanism in confronting the stressor or problem. However, another study found that such an adaptive coping mechanism previously applied will be transformed into a maladaptive coping mechanism when the student experiences an acute bullying situation (Clarke, 2009). Therefore, the coping mechanism is considered to be very important. Failure in developing the adaptive coping that responds to any presence of stressors often has an impact on students' health, welfare, and academic achievement (Deasy et al., 2014).

In this regard, this research is intended to identify the correlation between bullying and coping mechanisms during nursing students' clinical practices.

## 2 METHOD

### 2.1 *Study design*

This study uses a correlation analytic design with the cross-sectional approach. The correlation between bullying and the coping mechanisms of nursing practice students is analysed.

### 2.2 *Sample and sampling technique*

The sample selection method used in this research is non-probability purposive sampling. There were 138 nursing students at Andalas University who met the criteria for becoming respondents.

### 2.3 *Data c and analysis*

The data collection tool consisted of a questionnaire containing statements which were directly filled in by the students. These statements elicited the students' demographic data, bullying occurrence during the nursing students' professional practice, and their coping mechanisms.

## 3 RESULTS

The analysis shows that as many as 68.1% of the students perceived low level bullying events, 29.7% experienced moderate bullying, and 2.2% faced severe bullying occurrences.

The majority of students (94.9%) affirmed that the bullies were nurses. In terms of their response to bullying, most students (79.0%) had applied an adaptive coping mechanism. A significant correlation between bullying and the nursing students' coping mechanisms was obtained as a further result.

## 4 DISCUSSION

Bullying experienced by students can originate from a number of sources. This research found that the dominant bullying case experienced by the students was perpetrated by nurses (94.9%), followed by lecturers (53.6%), patients or their families (48.6%), classmates (37.7%), doctors (22.5%), and clinical counsellors (13.8%). This trend might occur because, during the process of the clinical practice, the students were inclined to associate with nurses and lecturers to obtain more guidance, duties, or work from them so nurses and lecturers became the major bullying sources.

This finding is in parallel with Palaz's (2013) study of nursing students in Turkey, where the nurses and lecturers (70.8% and 29.5% respectively) became the largest sources of bullying experienced by the students. Likewise, another study (Rahman, 2014) reported that nurses and lecturers became the leading bullies for nursing students at the university because the nursing students dealt with them more often while carrying out nursing practice. Moreover, the nurses regarded the nursing students as junior nurses who were subject to their instruction to execute a variety of tasks such as sending and picking up lab results as well as delivering the patients to the radiology room (Rahman, 2014; Kassem et al., 2015).

The results of the research showed that 92 (97.9%) of94 students who experienced low level bullying used adaptive coping and two students (2.1%) used a maladaptive coping mechanism. In the meantime, as many as 17 students (38.6%) who experienced moderate or severebullying utilised an adaptive coping mechanism and 27 students (61.4%) applied maladaptive coping.

The results indicate that there is a correlation between bullying and the coping mechanism used by the nursing practice students of Andalas University in 2016. This output is in line with Clarke (2009) whose research was conducted on nursing students in Canada. Clarke found that the more the students faced bullying, the more they would use a maladaptive coping mechanism such as self-blame, behaviour dismissal, emotional release, and self-diverting.

The study also indicated the tendency of the students who experienced moderate or severe bullying to utilising a maladaptive coping mechanism compared with those who encounter low level bullying. The maladaptive coping mechanism is a coping mechanism that hinders the function of integration, breaks growth, lowers autonomy, and has a tendency to dominate and control the surroundings (Stuart, 2013).

The coping mechanism that was executed by the students who encountered bullying might be varied. The different uses of coping mechanisms could be caused by the students' unequal coping skills. Students who experienced low level bullying were more able to adapt to the bullying problems than those who encountered moderate or severe bullying, who were more liable to use maladaptive coping mechanisms. Furthermore, the coping mechanism that was applied by each individual was not always the same because the individual's coping ability could be affected by several factors, such as physical health, culture, age, conclusiveness or positive consideration, problem-solving skills, social skills, spiritual and social support (Stuart & Laraia, 2005; Viedebeck, 2008).

The maladaptive coping mechanisms often used by the students are in form of avoiding the incoming problems, venting their emotions by saying or doing something bad, refusing to believe in what is going on, blaming themselves for what is happening, and intending to give up to confront that problems. All maladaptive coping mechanisms will not be able to solve the difficulties well. Instead, they will add to the difficulties, conflicts, tension, fear, and anxiety (Kartono, 2011).

Students who have used adaptive coping mechanisms tend to be open about their problems, take active steps, and think about the best way to deal with the problem. They also seek support and advice from others. The adaptive coping mechanism is a coping mechanism which supports the functions of integration, growth, learning, and achieving goals. When an individual uses adaptive coping, the negative effects that arise from a problem can be lessened (Stuart, 2013; Donoghue et al., 2014). Failure in developing the adaptive coping as the response to any appearing *stressor* often evokes an impact on the students' health, welfare, and academic merit (Deasy et al., 2014).

## 5 CONCLUSIONS

The researchers obtained the following conclusions. The students conducting professional practice who experienced bullying of a low level category were of a higher percentage compared to those who experienced bullying of moderate or high level categories. The students performing clinical practice who used adaptive coping mechanisms were of a higher percentage than those who used maladaptive coping mechanisms. There was a significant correlation between the level of bullying and students' coping mechanisms.

It is expected that students are capable of endeavouring to prevent occurrences of bullying by improving their knowledge and skills during practice along with improving the ability to socialise and communicate effectively in dealing with others. It is suggested that educational institutions create an educational environment that is free from bullying by providing a sense of security and comfort for students so that the process of accomplishing the education goal goes well. Institutions are also recommended to educate students about bullying behaviours and how to deal with bullying issues appropriately. It is expected that nurses and lecturers can review workloads or tasks that are assigned to the students, and it is preferable that the workloads or the tasks should be given at appropriate times. It is hoped that the next researchers will implement further investigations by adding other variables which are related to bullying or by doing further studies on bullying and students' coping mechanisms using a qualitative method so that the results of the research will be more significant and a deeper understanding about the problem of bullying can be acquired

## REFERENCES


Chakrawati, F. (2015). *Bullying siapa takut?: Panduan untuk mengatasi bullying. Bullying? Who scared!: A guide to overcome bullying.* Solo, Tiga Ananda.

Clarke, C. (2009). *The effects of bullying behaviours on student nurses in the clinical setting* (Thesis) University of Windsor, Ontario, Canada.

Clarke, M.C., Deborah, J.K., Dale, L.R.& Kathryn, D.L. (2012). Bullying in undergraduate clinical nursing education. *Journal of Nursing Education, 51*(5), 1–9.

Cooper, J.R.M., Jean, W., Rebecca, A., Jennifer, C.R. & Mary, M.N. (2011). Student's perceptions of bullying behaviors by nursing faculty. *Issues in Educational Research, 21*(1), 1–21.

Deasy, C., Barry, C., Julie, P. & Didier, J. (2014). Psychological distress and coping among higher education students: A mixed method enquiry. *Public Library of Science* [Online], *6*(12), 115–193. Available from: https://doi.org/10.1371/journal.pone.0115193. Accessed May 2016.

Donoghue, C., Angela, A., David, B., Gabriela, R.,& Ian, C.(2014). Coping with verbal and social bullying in Middle School. *International Journal of Emotional Education, 6*(2), 40–53.

Kartono, K. (2011) *Psikologi anakChildren psychology*.Mandar Maj:Bandung.

Kassem, A.H., Reda, S.E.& Wessam, A.E.(2015). Bullying behaviours and self efficacy among nursing students at clinical settings: Comparative Study. *Journal of Education and Practice*, *6*(35), 25–36.

Khater, W.A., Laila, M. Z., Insaf, A.S. (2014). Sources of Stress and Coping Behaviours in Clinical Practice among Baccalaurate Nursing Student. *International Journal of Humanities and Social Science, Vol. 4, No. 6*, 194–202.

Muhith, A. (2015). *Psychiatric Nursing Education.* Yogyakarta: ANDI.

Nursalam & Ninuk, D.K. (2011). The care of nursing in patients infected with HIV/AIDS. Jakarta: Salemba Medika.

Nursalam & Ferry, E. (2008). *Pendidikan dalam keperawatanEducation in nursing*. Jakarta, Salemba Medika.

Nursalam. (2008) *Konsep dan penerapan metodelogi penelitian ilmu keperawatan: pedoman skripsi, tesis, dan instrumen penelitian keperawatan. Concept and application of nursing research method: thesis guidance, nursing research instrument* Jakarta, Salemba Medika.

Palaz, S. (2013). Vertical bullying in nursing education: Coping behaviors of Turkish students. *International Journal of Nursing Education, 5*(1), 193–197.

Rahman, R.M.A.E. (2014). Perception of student nurses' bullying behaviors and coping strategies used in clinical settings. *Nursing Education Research Conference*. http://hdl.handle.net/10755/316820.

Seibel, M. (2014). For us or against us? Perceptions of faculty bullying of students during undergraduate nursing education clinical experiences. *Nurse Education in Practice, 14*(3), 271–274.

Sejiwa. (2008). *Bullying: Mengatasi kekerasan di sekolah dan lingkungan sekitar anak Bullying: Solving the violence in school and children environment.* Jakarta, Grasindo.

Smith, M.J., Roger, D.C., & Joyce, J.F. (2015). *Encyclopedia of nursing education.* New York: Springer Publishing Company.

Strauss, S. (2012). *Sexual harassment and bullying*. United Kingdom, Rowman & Littlefield.

Stuart, G.W. & Laraia, M.T. (2005). *Principles and practice of psychiatric nursing* (8th Ed.). St. Louis: Elsevier Mosby.

Stuart, G.W. (2013). *Principles and practice of psychiatric nursing* (10th Ed.). St. Louis: Elsevier Mosby.

Viedebeck, S.L. (2008). *Buku ajar keperawatan jiwa. Psychiatric nursing handbook* Jakarta, Publisher of medical book EGC.

Werner, S. (2012). *In safe hands: Bullying prevention with compassion for all.* United Kingdom, Rowman & Littlefield.

# *Maternal health*

*Strengthening Research Capacity and Disseminating New Findings in Nursing and Public Health – Malini et al. (Eds)*
*© 2018 Taylor & Francis Group, London, ISBN 978-1-138-50066-2*

# Promoting kangaroo mother care practice at home through educational video and nurse-assisted practice

N. Fajri
*Syiah Kuala University, Banda Aceh, Indonesia*

Y. Rustina & E. Syahreni
*University of Indonesia, Depok, Indonesia*

ABSTRACT: This study aims to identify the effect of Kangaroo Mother Care (KMC) education on maternal motivation and its practice at home. The study employed a quasi-experimental design with a consecutive sampling technique. It involved 32 mothers of infants with low birth weight, with 16 mothers in an intervention group that attended KMC education. The motivation of the two groups was calculated when the infants were leaving the hospital, while the implementation of KMC at home was measured after about three to four days at home. Data was analysed with a chi-square. The results showed that there was no difference observed in the motivation. However, an apparent difference is discerned in the KMC implementation between the two groups. It can be concluded that KMC education influences the implementation of KMC at home. It is suggested that hospitals introduce mothers to KMC through education, discussion, and real-life practice in order to improve the health of infants with a low birth weight.

## 1 INTRODUCTION

Low birth weight infants are the leading cause of neonatal mortality at arate of 35% each year (World Health Organization, 2012). This is because they are very prone to hypothermia, and even though hypothermia is rarely the direct cause of death, it is the most influencing factor in worsening infant conditions, such as on neonatal infections, prematurity, and asphyxia (Lunze et al., 2013). Quality care, like Kangaroo Mother Care (KMC), is very much required to reduce the morbidity and mortality rate.

KMC utilises direct skin contact between mothers and infants. Such contact has a significant impact on their survival, healthy growth and development. It can reduce the neonatal mortality rate (Engmann et al., 2013; Lawn et al., 2010). It is also very effective at reducing severe morbidity (especially from infections (Lawn et al., 2010)), can accelerate the infant's weight gain, prevent hypothermia, maintain the infant's physical stability (Rodriguez et al., 2007), reduce pain (Padhi et al., 2015), promote a longerdeepsleep, and reduce neonatal stress (Mörelius et al., 2015). Infants should receive these benefits not only at the hospital, but also at home.

Hypothermia is common in hospitals (32% to 85%) and at home (11% to 92%) (Lunze et al., 2013). Thus, mothers and other family members need to be prepared and trained to properly perform KMC and continue to do so at home. The implementation of KMC at home requires strong motivation from both mothers and families, and strong support from the community.

Many benefits are associated with KMC, but few neonates receive such benefits (Engmann et al., 2013). Amongst the many factors which contribute to this situation are the knowledge of the mothers and other family members, support the mothers receive to implement KMC (especially at home), lack of knowledge, attitude, skills of the health workers, and policies and facilities in the local health services (Pratomo et al., 2012). In addition to the mother's knowledge, the implementation of KMC may also be affected by the mother's motivation to implement KMC since motivation, understanding and action are undeniably interconnected. Motivation can encourage and strengthen a person to act (Redman, 1993). There have been some studies on the implementation of KMC at home, but no studies were found that examined the education of mothers using videos, their motivation and the sustainability of KMC practice at home. Therefore, this study seeks to learn the influence of KMC education on the motivation and KMC practice at home.

## 2 METHOD

### 2.1 *Study design*

The study used a quasi-experimental design with a post-test only nonequivalent control group approach.

### 2.2 *Technique of sampling*

Consecutive sampling was used to draw 32 mothers with low birth weight infants from those that were being treated in a perinatology room and were in a stable condition. A total of 16 mothers were placed in the intervention group and 16 mothers in the control group.

### 2.3 *Ethical considerations*

The ethical study was completed by the Research Ethics Committee of the Nursing Faculty at the University of Indonesia.

### 2.4 *Data collection and analysis*

The intervention group attended KMC education (educational video, discussion and practice) about two to four days before leaving the hospital. The control group attended hospital-standard KMC education at the discharge time (without media and practice). Data was obtained from a post-test that measured the effect of education on the mothers' motivation to perform KMC at home; the measurement was undertaken when the mothers were leaving the hospital. In addition, the implementation of KMC at home was measured when the infants were in the third and fourth day at home.

In order to compare the effect of education, the author used a control group who received standard treatment from the hospital. An instrument consisting of 26 items, made subjectively by the author, was used to measure the motivation with the Cronbach alpha of 0.877. The data were then analysed using a computer program.

## 3 RESULT

The majority of the respondents were in their early adulthood (62.5%) and only 3.1% of the respondents were adolescents. The rest of the respondents were in their mid-adulthood (34.4%). All respondents were high school and university graduates and no respondents were elementary school graduates. The majority of the respondents were university graduates (68.8%), while high school graduates occupied 31.3% of the sample. In terms of emotional readiness, 50% of the respondents were emotionally ready while the rest, 50%, were not ready to perform KMC at home.

In the intervention group, the majority of the mothers took part in the educational video without any family assistance (62.5%), while 34.4% had family assistance. In the hospital, 62.5% of the mothers performed KMC two times while the rest (37.5%) performed KMC less than two times. The following is the results of the bivariate analysis:

Table 1. Motivational difference of KMC practice on mothers who had low birth weight infants (based on groups).

| | Motivation | | | | | | | | |
|---|---|---|---|---|---|---|---|---|---|
| | Low | | High | | Total | | | | |
| Groups | n | % | n | % | n | % | *OR* (95% CI) | $\chi^2$ | *p*-value |
| Control | 8 | 50 | 8 | 50 | 16 | 100 | 1 | | |
| Intervention | 7 | 43.8 | 9 | 56.3 | 16 | 100 | 1.28 | 0.001 | 1 |
| Total | 15 | 46.9 | 17 | 53.1 | 32 | 100 | (0.32–5.16) | | |

(OR = Odds Ratio, CI = Confidence Interval).

Table 2. Difference of KMC practice at home on mothers with low birth weight infants (according to groups).

| | Implementation of KMC at home | | | | | | | | |
|---|---|---|---|---|---|---|---|---|---|
| | Not implemented | | Implemented | | Total | | | | |
| Groups | n | % | n | % | n | % | *OR* (95% *CI*) | $\chi^2$ | *p*-value |
| Control | 10 | 62.5 | 6 | 37.5 | 16 | 100 | | | |
| Intervention | 0 | 0 | 16 | 100 | 16 | 100 | – | 11.78 | 0.001 |
| Total | 10 | 31.3 | 22 | 68.8 | 32 | 100 | | | |

(OR = Odds Ratio, CI = Confidence Interval).

Table 1 shows that 50% of the mothers in the control group were highly motivated to perform KMC at home, while those in the intervention reached 56.3%. The statistical test indicated that there is no statistical difference in terms of motivation to perform KMC at home between the control and intervention group (*p*-value > 0.05).

Table 2 shows that 37.5% of the respondents in the control group performed KMC at home to their infants, while all respondents (100%) in the intervention group performed KMC at home. Statistically, there is a significant difference in term of KMC implementation between the control and intervention groups with $p < 0.05$.

## 4 DISCUSSION

The majority of the mothers in the intervention group were highly motivated to perform KMC. Looking at the figures of the two groups, more mothers in the intervention group were highly motivated to perform KMC at home compared to the control group (with a 6.3% discrepancy between the two groups). At such a very low difference, statistically, it showed that there is no significant difference in terms of motivation between the two groups (*p*-value = 1.00).

According to Bastable (2002), motivation is influenced by the cognitive state of a person. Both groups were given KMC education in different ways, but as both groups received information which enhanced their knowledge on KMC, all of them were motivated to perform KMC at home.

A study by Muddu and Boju (2013) proposed that mothers can easily understand KMC education if delivered clearly and simply, including with the use of local language. 95% of the mothers in the study had positive feelings towards implementing KMC at home and their enthusiasm to perform KMC at home increased after participating in a simple KMC education programme. Furthermore, KMC practice can lower anxiety for mothers as found by Saidah et al. (2011). In this study, mothers with Low Birth Weight LBW infants who received standard treatment, also attended KMC education that was delivered upon leaving the hospital. After they listened to the standard explanation of KMC, their motivation to perform KMC at home was promptly measured. There was no time distance between the standard education treatment and motivation measurement. From the results, the information they received at the hospital elevated their motivation to perform KMC at home.

Generally, *Acehnese* people consider Muslim scholars or religious leaders as role models in making a decision(Wahyuningroem, 2005). Muslim scholars are not only referred to for religious affairs, but also for education and family issues. Probably, the involvement of Muslim scholars in promoting KMC practice at home can increase a mother's motivation to perform KMC at home. Furthermore, the mothers of the mothers with LBW infants played a major role in making a decision regarding infant care at home. Therefore, family support is very important in the implementation of KMC (Pratomo et al., 2012).

The provision of educational videos, discussion and the practising of KMC in the hospital significantly enhanced the implementation of KMC at home. The use of videos as educational media can stimulate many senses at once, such as visual and aural, which eventually stimulateseagerness to perform KMC at home. According to a study byYazar and Arifoglu (2012), educational videos can have a positive effect and improve creativity compared to education with static media and education without the use of media at all.

This practice is in accordance with the principles of KMC education that utilises counselling principles, since the problems of LBW infants vary, and the responses of their parents towards the problems they and their infants face are also diverse (Perinasia, 2011).

In this study, due to various reasons, no mothers with LBW infants could simultaneously perform KMC at home. Amongst the reported problems were the inability of an infant to withstand being in a kangaroo sling for a long time during hot weather, the need for the mothers to do housework, and mothers alone at home with their infant while their husbands were at work. Some mothers needed to take care of their other children and could not perform KMC for 24 hours a day. They were also concerned that if they had to do it at night, they could not control the position of their infants.

Similar results were reported in study by Blomqvist et al. (2013) in that one of the problems restricting the mothers from continuously practising KMC was the uncomfortable position of KMC during sleep, which lead to sleep deprivation. The continuous implementation of KMC requires cooperation from and knowledge of the nurses/health workers to deliver and convince the mothers of LBW infants, and the community, about the importance of continuous KMC at home.

Studies on the resistance towards KMC practice conducted in 15 developing countries found that one of the causes was the lack of confidence in the need for continuous KMC. In order to be able to convince the family, health workers must first be definitely assured about the information they are about to share. Very premature infants do not have the ability to regulate their body temperature well, so the non-continuous implementation of KMC can lead to growth disorders, hypothermia, and complications, such as hypoglycaemia and apnoea (Charpak & Ruiz-Peláez, 2006).

There was no significant difference between the two groups in terms of the KMC implementation at home by mothers who were highly motivated. The majority of the mothers who werehighly motivated carried out KMC at home.

The study conducted by Joussemet et al. (2014) showed that a child's condition also influenced their parents' motivation in caring for them. This follows the results of this study and is especially true for the mothers of LBW infants. Parents with LBW infants were found to be motivated to take care of their infants in the best way and perform the treatment at home.

In contrast, the results of the study with low motivation carried out by Carter et al. (2002) suggested that motivation cannot be a significant measurement to predict health behaviours, its measurement cannot be calculated effectively and it has a less clear concept. Motivation still requires a clear definition before the measurement can be carried out.

The majority of the mothers who performed KMC in the hospitals were more likely to have a higher motivation to perform it again at home. This is in line with the suggestions by Blomqvist et al. (2012) that mothers who practised KMC were more motivated to continue practising KMC on their infants because they could feel the benefits for their infants. Similar results were also observed in the intervention group of this study who had practised KMC in the hospital. They were highly motivated to continue KMC at home and continued to do so at home.

This study shows that the frequency of KMC practice during the hospital stay varied ranging from one to four times of each LBW infant. Health workers need to support and motivate mothers who visit the perinatology room to perform KMC. It was found that mothers often face many obstacles which hamper the implementation of KMC, such as having enough spare time to care for other family members at home and doing other domestic activities, and a feeling that it is not much a mother can do when she visits her sick child in hospital. Blomqvist et al. (2013) stated that families cannot freely and regularly be in the Neonatal Intensive Care Unit (NICU) room, especially during the nursing round, because nurses have to maintain the confidentiality of other patients; this, in turn, may inhibit the implementation of KMC. Furthermore, Blomqvist et al. (2013) said that some mothers had difficulty in dividing their time between visiting their infants treated at the NICU and being with other family members at home, especially with the siblings of the LBW infants.

All mothers of LBW infants who had watched educational videos, became involved in discussions and performed KMC in the hospital, continued to carry out KMC at home despite whether their family was present during the education session and irrespective of the number of KMCs performed during their time in the hospital. The duration of KMC practice at home varied and no respondents performed KMC continuously at home due to various problems they faced. The results of the study by Blomqvist et al. (2013) showed that mothers have more time to perform KMC in the hospital instead of at home. In addition to doing KMC, mothers also have to do other housework duties, such as shopping, cooking, washing clothes and welcoming guests.

The limitation of the current study was that there were some respondents who had no other family members at home during the visit, making it difficult to confirm the truth whether or not KMC was carried out at home. In addition, two hospitals in this study did not have a special room for watching videos and conducting discussions. As for the results, the treatments were carried out under limited conditions.

## 5 CONCLUSION

There was no significant difference in terms of motivation to implement KMC at home between the mothers who were given educational videos, discussions and practised KMC in the hospital, and those who received the standard education.

There was a significant difference in the implementation of KMC at home between the mothers who were given educational videos, discussions and practised KMC in the hospital, and those who received the standard education. Therefore, it can be concluded that KMC education influences the implementation of KMC at home on the mothers with LBW infants.

It is suggested that health services in hospitals provide educational videos, discussions and real-time practice of KMC in hospitals to mothers, and their families, of LBW infants in order to avoid hypothermia and improve the health of their infants.

## REFERENCES

Blomqvist, Y.T., Frölund, L., Rubertsson, C. & Nyqvist, K.H. (2013). Provision of kangaroo mother care: Supportive factors and barriers perceived by parents. *Scandinavian Journal of Caring Sciences*, *27*(7), 345–353. doi.org/10.1111/j.1471-6712.2012.01040.

Blomqvist, Y.T., Rubertsson, C., Kylberg, E., Jöreskog, K. & Nyqvist, K.H. (2012). Kangaroo mother care helps fathers of preterm infants gain confidence in the paternal role. *Journal of Advanced Nursing*, *68*, 1988–1996. doi.org/10.1111/j.1365-2648.2011.05886.

Carter, K.F., Kulbok, P.A. & Carter, K. (2002). Motivation for health behaviours: A systematic review of the nursing literature. *Journal of Advanced Nursing*, *40*(3), 316–330.

Charpak, N. & Ruiz-Peláez, J.G. (2006). Resistance to implementing kangaroo mother care in developing countries, and proposed solutions. *Acta Paediatrica*, *95*(46), 529–534. doi.org/10.1080/08035250600599735.

Engmann, C., Wall, S., Darmstadt, G., Valsangkar, B. & Claeson, M. (2013). Consensus on kangaroo mother care acceleration. *The Lancet*, *382*(13), 26–27. doi.org/10.1016/S0140-6736(13)62293-X.

Joussemet, M., Rene, T.J. & Koestner, R. (2014). Autonomous and controlled motivation for parenting: Associations with parent and child outcomes. *Journal of Child and Family Studies*, *24*(7), 1932–1942. doi.org/10.1007/s10826-014-9993-5.

Lawn, J.E., Mwansa-Kambafwile, J., Horta, B.L., Barros, F.C. & Cousens, S. (2010). Kangaroo mother care to prevent neonatal deaths due to preterm birth complications [Abstract]. *International Journal of Epidemiology*, i144–i154. doi.org/10.1093/ije/dyq031.

Lunze, K., Bloom, D.E., Jamison, D.T. & Hamer, D.H. (2013). The global burden of neonatal hypothermia: Systematic review of a major challenge for newborn survival. *BMC Medicine*, *11*(1), 24. doi.org/10.1186/1741-7015-11-24.

Mörelius, E., Örtenstrand, A., Theodorsson, E. & Frostell, A. (2015). A randomised trial of continuous skin-to-skin contact after preterm birth and the effects on salivary cortisol, parental stress, depression, and breastfeeding. *Early Human Development*, *91*(1), 63–70. doi.org/10.1016/j.earlhumdev.2014.12.005.

Muddu, G.K. & Boju, S.L. (2013). Knowledge and awareness about benefits of kangaroo mother care. *Indian Journal of Pediatrics*, *80*(10), 799–803. doi.org/10.1007/s12098-013-1073-0.

Padhi, T.R., Sareen, D., Pradhan, L., Jalali, S., Sutar, S., Das, T., ... Behera, U.C. (2015). Evaluation of retinopathy of prematurity screening in reverse kangaroo mother care: A pilot study. *Eye*, *29*(4), 505–508. doi.org/10.1038/eye.2014.340.

Perinasia. (2011). *Kangaroo mother care*. Jakarta, Indonesia: Perinasia.

Pratomo, H., Uhudiyah, U., Poernomo, I., Sidi, S., Rustina, Y., Suradi, R., ... Gipson, R. 2012. Supporting factors and barriers in implementing kangaroo mother care in Indonesia. *Paediatrica Indonesiana*, *52*(1), 43–50.

Redman, B.K. (1993). *The process of patient education*. St. Louis: Mosby-Year Book.

Rodriguez, A., Nel, M., Dippenaar, H. & Prinsloo, E. (2007). Good short-term outcomes of kangaroo mother care in infants with a low birth weight in a rural South African hospital. *South African Family Practice*, 49, 15–15c. doi.org/10.1080/20786204.2007.10873550.

Saidah, Q.I., Rustina, Y. & Nurhaeni, N. (2011). Decreasing maternal anxiety and improved wake-up status of low birth weight through treatment of kangaroo method.*Jurnal Keperawatan Indonesia*, *14*(3), 193–198.

Wahyuningroem, S.L. (2005). The role of women and the new era in Nanggroe Aceh Darussalam. *Antropologi Indonesia*, *29*(1), 93–101.

World Health Organization. 2012. *Born too soon: The global action report on preterm birth*. Geneva, Austria: WHO Press.

Yazar, T. & Arifoglu, G. (2012). A research of audio visual educational aids on the creativity levels of 4–14 year old children as a process in primary education. *Procedia—Social and Behavioral Sciences*, *51*, 301–306. doi.org/10.1016/j.sbspro.2012.08.163.

*Strengthening Research Capacity and Disseminating New Findings in Nursing and Public Health – Malini et al. (Eds)*
*© 2018 Taylor & Francis Group, London, ISBN 978-1-138-50066-2*

# An ergonomic breastfeeding chair design

P. Fithri, L. Susanti, I. Arief, D. Meilani & C.T. Angelia
*Faculty of Engineering, University of Andalas, Kampus Limau Manis, Pauh, Padang, West Sumatra, Indonesia*

ABSTRACT: Breast milk plays an important role in a baby's brain development and disease prevention. Indonesian government regulations require mothers to breastfeed their babies for at least the first six months of its life. There is however, a general lack of proper breastfeeding facilities in public nursery rooms. Stiffness and pain in the back, neck, legs, and arms of breastfeeding mothers is commonly found due to poorly ergonomically designed facilities. Therefore, it is necessary to design a facility for breastfeeding mothers such as an ergonomic chair. The research for the design of such a chair started with the collection of anthropometric data to determine suitable chair dimensions, and then the application of the Quality Function Deployment (QFD) method to identify priority features that should be put into the design. The final designs had adjustable armrests, a sliding footrest, a baby support cushion, a headrest, and an adjustable backrest.

## 1 INTRODUCTION

Breast milk (or a 'mother's milk') has very important health benefits for the baby (Agostoni. C, et al., 2009). Besides functioning as a perfect food to help the baby's growth process, breast milk also helps increase the baby's resistance to various infectious bacteria, fungi, and viruses (Anatolitou, 2012). Due to the benefits of the mother's milk, the government of the Republic of Indonesia follows the World Health Organisation's recommendation for a mother to breastfeed a child for at least the first six months of its life to maintain the baby's health.

There are several obstacles in implementing a government regulation on mandatory exclusive breastfeeding. One of them is inadequate facilities, especially the unavailability of seats in lactation rooms. It is common that seats available in the lactation rooms, such as in the offices, business centres or even at home, do not respond to user's complaints that they are inconvenient. The only facility commonly provided are chairs which do not reflect the user's specific needs. While breastfeeding, such body parts as the back, shoulders, neck, hands and feet tend to become uncomfortable. This is because the seats do not consider the special needs of the mothers and consequently, the body position while breastfeeding will be wrong.

In accordance with the ergonomic principle of 'fitting the task to the man rather than the man to the task' (Tarwaka, 2004), it is the facility that should be adjusted to the human limitations (Nurmianto, 2004), not the other way around. Hence, the solution offered is a chair that can accommodate breastfeeding mothers through biomechanics and an anthropometric approach to reduce muscle soreness when breastfeeding.

The problem to be resolved in this study is concerned with how to design an ergonomic chair for breastfeeding mothers in accordance with their needs both in terms of dimensions and comfort, while keeping it within relatively affordable prices.

## 2 METHODS

### 2.1 *Identify customers' needs*

Customers' needs were identified through a survey on 100 nursery women. The identified needs of the breastfeeding mothers to be addressed were as follows: strong seat to support the mother and the baby; armrests or other supports on the side of the chair; a footrest; the angle of the chair's back to be comfortable for the user; comfortable seat design; a cushion to support the baby; a headrest; and moveable seat.

### 2.2 *Anthropometric data*

The anthropometric data collection involved 30 respondents. Predetermined anthropometric variables used in this study were popliteal height, popliteal buttocks, hip width, shoulder width, shoulder height, elbow height, leg width, and leg length. Data was collected directly by anthropometric measurement of the breastfeeding mothers. 30 data sets were collected and tested using the normality test to see whether the data could be used as a reference to determine the dimensions of the chair parts.

### 2.3 *Quality function deployment method*

The Quality Function Deployment (QFD) method is used to determine the characteristics that designers must meet in order to produce customer-oriented products (Cohen, 1995). In this research, the development phase of the House of Quality (HOQ) matrix consisted of several stages:

1. Creating a list of breastfeeding mothers' needs for the products to be designed (customers' requirements).
2. Determining the technical characteristics (technical response) to meet the breastfeeding mothers' needs. In this section, the designer translates what is needed and defines technical characteristics to meet the needs of the consumers. This is then validated directly by the experts in the field of production.
3. Determining the relationship between breastfeeding mothers' needs and the technical characteristics required in order to see the relationship between each of the existing characteristics. Using a strong correlation technique means that if one of the characteristics of the technique is not a match, it will significantly affect the characteristics of the other techniques.
4. Determining the relationship among all technical characteristics.
5. Creating a Phase 2 HOQ matrix that contains part characteristics or detailed technical characteristics. This phase is a continuation of the technical response to a Phase 1 HOQ matrix. In a Phase 2, the technical response is detailed by the expert in the field of production and becomes part of the characteristics that facilitates the developer in the product design stage. The stages are the same as in a Phase 1 HOQ matrix, where percentages and priorities for each part characteristic to be applied in the design are searched, which ultimately strengthens the relationship between the technical response and part characteristics.

## 3 PRODUCT DESIGN

### 3.1 *Adjustable armrest*

The armrest is used to support the mothers' elbows and hands while feeding her child. The presence of an armrest can reduce fatigue on the hands.

### 3.2 *Sliding footrest*

A footrest places the knee slightly higher or equal to the pelvis. This position allows the breastfeeding mother to feel comfortable and reduce fatigue while sitting because the legs are not hanging nor placed on the floor.

The design of the ergonomic breastfeeding chair in this study features a sliding footrest which is not separate from the seat. The sliding is able to change the buffer position by pulling and extending it.

The footrest slider has rails on each side of the chair to pull and lengthen the footrest. The maximum reach of the slider withdrawal was designed to be 40 cm in order to meet the maximum percentile of respondents. An image of the sliding footrest used in this study is shown in Figure 1a. The footrest has a flexible angle that can be changed as per the user's convenience.

### 3.3 *Support for the baby*

Breast pads are often used by mothers to support their sleeping babies while breastfeeding and the pads are separate from the chair. The breastfeeding chair in this study was designed to have a straightforward indivisible pad which negated the need to add another cushion to support the baby. The design considered the thick anthropometry of the stomach so that the nursing mother could comfortably use a breastfeeding pillow. The thickness of the breastfeeding pad used was 5 cm, with a soft foam type. This was based on a recommendation for a 3.8 cm thick 'comfortable' foam (Perwita, 2009) which offers a high level of comfort that is required for a newborn.

### 3.4 *Headrest*

The survey within this research revealed that some respondents wanted headrests to be more comfortable. The chair was designed with a headrest, which consisted of two designs: directly attached to the seat frame and cannot be removed; and made separately from the seat frame so that it can be removed (because not all respondents considered the headrest as important). The second design, nevertheless, was technically more difficult when compared to the first headrest design.

### 3.5 *Wheels*

Some respondents wanted a breastfeeding chair equipped with wheels for easy movement. The type of wheel used in the chair design for this study had a lock to keep the chair immobile while being used.

### 3.6 *Adjustable backrest tilt*

According to the American National Standards Institute (ANSI) in Perwita (2009), the recommended backrest angle for breastfeeding mothers is105°. However, some respondents to this study considered that an adjustable backrest angle should be included in the chair design. Such an adjustable feature certainly adds a high value of comfort to the users while feeding

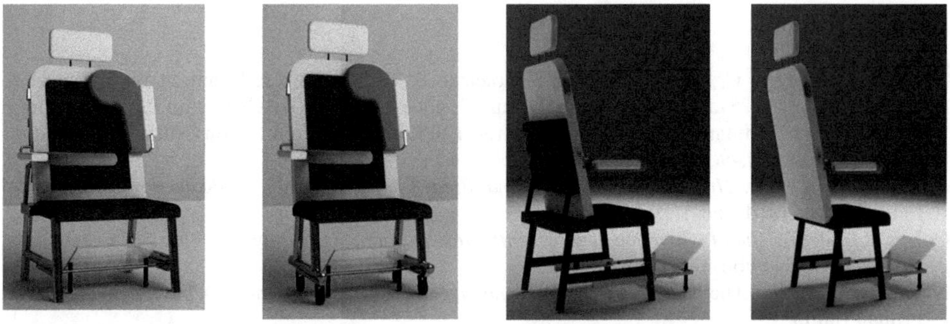

(a) (b) (c) (d)

Figure 1. Designs 1 to 4 of the proposed ergonomic breastfeeding chair.

The backrest of the chair in this study was based on the ANSI recommendations, but was also designed to be manually adjustable into several angles of 95°, 105° and 110°.

### 3.7 *Design specification for an ergonomic breastfeeding chair*

The design of an ergonomic chair for this study was based on the dimensional calculations and a determination of the required features. The chair design featured an adjustable armrest, a sliding footrest, was foam coated, and contained seat cushions that could support the baby during nursing. However, additional and alternative features were also considered. This culminated in the design of four slightly different ergonomic breastfeeding chairs, and these are discussed further in paras 3.7.1–3.7.4.

#### 3.7.1 *Design 1*

Design 1 was constructed with reference to predetermined dimensions, predefined features, and added features, such as a head cushion attached directly to the seat frame, and an adjustable angle for the backrest. Design 1 is presented in Figure 1(a).

#### 3.7.2 *Design 2*

Design 2 has many similarities to Design 1, but with different additional features, such as removable headrests and wheels. Design 2 is shown in Figure 1(b).

#### 3.7.3 *Design 3*

Design 3 has some additional features, such as an adjustable angle for the backrest and headrests that are directly embedded into the seat frame. Design 3 is shown in Figure 1(c).

#### 3.7.4 *Design 4*

Design 4 has similar material to Design 3, but had a removable headrest. Design 4 is shown in Figure 1(d).

## 4 CONCLUSIONS

The most important features of a breastfeeding chair that will meet the needs of breastfeeding mothers includes an adjustable armrest, a sliding footrest, a cushion to support the baby, a headrest, foam coating, an adjustable backrest angle, and wheels. The different designs shown and discussed in this study were found to meet the ergonomic needs of breastfeeding mothers because they took into account their anthropometric data and the biomechanical analysis using the Ovako Work Posture Analysis System method. This research can be further developed in the creation of prototypes and the evaluation of the technical details of each chair design to the stage of manufacturability.

## REFERENCES


Agostoni C, Braegger C, Decsi T, Kolacek S, Koletzko B, Michaelsen KF, Mihatsch W, Moreno LA, Puntis J, Shamir R, Szajewska H, Turck D, van Goudoever J; ESPGHAN Committee on Nutrition (2009). Breast feeding: A Commentary by the ESPGHAN Committee on Nutrition. J Pediatr Gastroenterol Nutr;49(1):112–25.

Anatolitou, F (2012). *Human milk Benefits and Breastfeeding*. Journal of Pediatric and Neonatal Individualized Medicine. 1(1). 11–18.

Cohen, L. (1995). *Quality function deployment: How to make QFD work for you*. Massachusetts: Addison-Wesley Publishing Company.

Nurmianto, E. (2004). The *Basic Concept of Ergonomics and Applications* (2nd ed.). Surabaya, Indonesia: Prima Printing.

Perwita, I (2009). *Proposed Design of Breastfeeding Chair*. Surakarta: Universitas Sebelas Maret.

Tarwaka., Solichul, H.A. Bakri. & Sudiajeng, L. (2004). *Ergonomics for occupational safety and health and productivity (1st ed.)*. Surakarta, Indonesia: UNIBA Press.

*Strengthening Research Capacity and Disseminating New Findings in Nursing and Public Health – Malini et al. (Eds)*
*© 2018 Taylor & Francis Group, London, ISBN 978-1-138-50066-2*

# Minangkabau mothers' characteristics and breastfeeding self-efficacy in relation to exclusive breastfeeding practice in Padang

V. Priscilla
*University of Andalas, Padang, Indonesia*

ABSTRACT: A mother's breastfeeding self-efficacy and characteristics affect breastfeeding activities and competency. This study aimed to determine the relationship between a mother's characteristics and self-efficacy with breastfeeding. This is a quantitative study using a descriptive analytic method with a cross sectional design. This study was conducted in Padang in 2016 with 388 participants. There are significant similarities between the characteristics of occupation, education, knowledge, mother's attitude, and the self-efficacy with breastfeeding ($p < 0.05$). Maternal characteristics and breastfeeding self-efficacy are modifiable determinants that health professionals need to observe in order to improve breastfeeding outcomes.

## 1 INTRODUCTION

Breastfeeding is the most effective way to ensure a baby's health and welfare. However, the facts show that less than 40% of babies under six months old in the world are given exclusively breast milk (WHO, 2015). In 2013, Indonesian exclusive breastfeeding attainment was only 54.3% and 55.7% in 2016. If we compare the attainment in 2013 and 2016, Indonesia has not accomplished the WHO's target of 1.2% annual escalation of breastfeeding attainment. Furthermore, breastfeeding attainment in West Sumatra had reached 68.9% in 2013 and 70% in 2015 while breastfeeding attainment in Padang was 72.83% (Indonesian Ministry of Health, 2016). One of the most crucial factors that can affect breastfeeding activity is breastfeeding self-efficacy. Demography and the characteristics of a mother cannot be denied as the determining factors of breastfeeding accomplishment.

## 2 RESULT

The majority of respondents (63.1%) in this study did not perform exclusive breastfeeding practice.

The exclusive breastfeeding rate was higher among mothers whose ages were in the range of 20–35 years old than those under 20 years old (63.2%) and older than 35 years old (42.8%) with $p = 0.081$.

Multiparous mothers (45%) had a higher exclusive breastfeeding rate if compared to primiparous (42.1%) and grand multiparity mothers (38.1%) with the significance value of $p = 0.051$.

Here, working mothers (60%) were in a higher rate in exclusive breastfeeding if compare to housewives (38%) with $p = 0.000$.

Highly educated mothers (59.5%) had a higher rate of exclusive breastfeeding than mothers with lower education level (41.9%) with $p = 0.045$.

The exclusive breastfeeding rate was higher among mother who have sufficient knowledge (56.3%) when compared with those less knowledgeable mothers (33.6%) with $p = 0{,}000$.

Mothers with positive attitudes appear to have a higher rate (54.6%) of exclusive breastfeeding than mothers with negative attitudes (34.1%) with $p = 0{,}000$.

Table 1. Distribution frequentation of Exclusive Breastfeeding Practice (EBP).

| Exclusive breastfeeding (0–6 months old) | Frequency | % |
|---|---|---|
| Exclusive | 143 | 36.9 |
| Non exclusive | 245 | 63.1 |
| Total | 388 | 100 |

Table 2. The relationship of a mother's characteristic: maternal age and exclusive breastfeeding practice.

| Maternal age (year) | Exclusive breastfeeding practice | | | |
|---|---|---|---|---|
| | Yes | No | Total | P value |
| ≥20–≤35 | 158 (42.8%) | 211 (57.2%) | 369 (100%) | 0.081 |
| <20 > 35 | 12 (63.2%) | 7 (36.8%) | 19 (100%) | |

Table 3. The relationship of a mother's characteristic: parity and exclusive breastfeeding practice.

| | Exclusive breastfeeding | | | |
|---|---|---|---|---|
| Parity | Yes | No | Total | P value |
| Primiparous | 72 (42.1%) | 99 (57.9%) | 171 (100%) | |
| Multiparous | 90 (45.9%) | 106 (54.1%) | 196 (100%) | 0.051 |
| Grand multiparous | 8 (38.1%) | 13 (61.9%) | 21 (100%) | |

Table 4. The relationship of a mother's characteristic: occupation and exclusive breastfeeding practice.

| | Exclusive breastfeeding practice | | | |
|---|---|---|---|---|
| Occupation | Yes | No | Total | P value |
| Housewife | 110 (38.2%) | 178 (40.5%) | 288 (61.8%) | 0,000 |
| Career | 60 (60%) | 40 (40%) | 100 (100%) | |

Table 5. The relationship of a mother's characteristic: education and exclusive breastfeeding practice.

| | Exclusive breastfeeding practice | | | |
|---|---|---|---|---|
| Education | Yes | No | Total | P value |
| Higher | 25 (59.5%) | 17 (40.5%) | 42 (100%) | 0.045 |
| Lower | 145 (41.9%) | 201 (58.1%) | 346 (100%) | |

Table 6. The relationship of a mother's characteristic: knowledge and exclusive breastfeeding practice.

| Mother's knowledge | Exclusive breastfeeding practice | | | |
|---|---|---|---|---|
| | Yes | No | Total | P value |
| Good | 98 (56.3%) | 76 (43.7%) | 174 (100%) | 0,000 |
| Poor | 72 (33.6%) | 142 (66.4%) | 214 (100%) | |

Table 7. The relationship of a mother's characteristic: attitude and exclusive breastfeeding practice.

| Mother's attitude | Exclusive breastfeeding Yes | No | Total | P value |
|---|---|---|---|---|
| Positive | 100 (54.6%) | 83 (45.4%) | 183 (100%) | 0,000 |
| Negative | 70 (34.1%) | 135 (65.9%) | 205 (100%) | |

Table 8. The relationship between breastfeeding self-efficacy and exclusive breastfeeding practice.

| Self-efficacy | Exclusive breastfeeding Yes | No | Total | P value |
|---|---|---|---|---|
| High | 116 (51.6%) | 109 (66.9%) | 225 (100%) | 0,000 |
| Low | 54 (33.1%) | 54 (33.1%) | 163 (100%) | |

Finally, the exclusive breastfeeding rate was higher among mothers with a high level of breastfeeding self-efficacy (51.6%) compared with those with a low level of breastfeeding self-efficacy (33.1%) with $p = 0.000$.

## 3 DISCUSSION

Minangkabau is one ethnic group that embraces the matrilineal system where the family line is based on the mother (Bhanbhro, 2017). The matrilineal kinship system provides an opportunity for mothers to have a prominent role especially in the *Rumah Gadang* (Minangkabau traditional house of extended families of a clan) setting, starting from arranging the family finances to the children's education. Minangkabau women not only take part in domestic but also their social environment matters. Bhanbhro (2017) explained that Minangkabau women value their significant roles in social and public life, especially during indigenous ceremonies and festivals. In addition to having a role in the *Rumah Gadang*, Minangkabau women have the opportunity to engage in economic activities and maintain a high enthusiasm to pursue higher levels of education. Since the 20th century, history has recorded that many Minangkabau women had attended formal or informal educational institutions (Yati et al., 2014). However, these abundant cultural roles have since become one of the barriers against the accomplishment of exclusive breastfeeding practice.

The results of this study show that exclusive breastfeeding in Padang, the capital city of Minangkabau province, West Sumatra, was relatively low (36.9%). The result confirmed Hafizan et al., (2014) findings in Malaysia where only 44.3% of the 159 respondents were exclusive breastfeeding. The relationship between maternal age and exclusive breastfeeding practice differs from place to place. This study shows that exclusive breastfeeding practice is not affected significantly by maternal age. Ogunlesi (2010) posited that maternal age is not a significant determinant of breastfeeding practice which p value $> 0.05$. This argument is supported by Hafizan et al., (2014) study in Malaysia showing that maternal age was not associated with duration and exclusive breastfeeding practice. Other studies have demonstrated that maternal age influenced breastfeeding initiation and duration. Ukegbu (2010) study shows that older maternal age is associated with exclusive breastfeeding and longer duration of breastfeeding which p value $< 0.05$. This finding suggests that the relationship between maternal age and exclusive breastfeeding practice varies from place to place. Therefore, health workers should understand how maternal age influences breastfeeding practice in order to plan a better intervention promotion.

This current study shows that parity is not significantly related with exclusive breastfeeding practice, which is in line with Emmanuel's (2015) finding. Some studies have shown that parity did not confer any advantage to breastfeeding practice, meaning that breastfeeding behaviour of primiparous and multiparous mothers is all the same. Hafizan et al., (2014) study on 159 mothers has shown that such sociodemographic factors as age, religion, and parity of the mothers were not associated with duration and exclusive breastfeeding practice.

This study has shown that occupation is significantly related with breastfeeding practice. Working mothers exclusively breastfed their babies in higher percentages than unemployed mothers. According to Danso (2014), around 90.5% of work status and 7.5% of family members' influence on exclusive breastfeeding negatively affect the efforts and decisions for professional working mothers to exclusively breastfeed their babies. On the other hand, professional working mothers tend to obtain adequate information on exclusive breastfeeding and its benefit for the babies and the mother themselves. Therefore, despite being employed, the working mothers can exclusively breastfeed their babies for a longer duration.

Education is one important factor that determines mothers' breastfeeding practice. This study has shown that the level of education is significantly related with exclusive breastfeeding practice. Mothers with high levels of education are more likely to breastfeed their infants for a longer period of time. Highly educated woman may be able to breastfeed exclusively as recommended because they are more likely to understand the benefits of breastfeeding when compared with less educated women who may breastfeed simply as a tradition and may not see any other need or benefit from that practice (Emmanuel, 2015).

This study, which confirmed Maonga, Mahande, Mamian, and Msuya's (2016) study, shows that maternal knowledge is significantly related with exclusive breastfeeding practice. According to Maonga et al., (2016), mothers with good knowledge of exclusive breastfeeding have higher odds of exclusively breastfeeding than others. Women who knew how long they should breastfeed a child showed a longer duration of exclusive breastfeeding and total breastfeeding than those who did not. Therefore, women who do not have the knowledge require counselling during antenatal care (Emmanuel, 2015).

In relation to mothers' attitude, this study has shown that the attitude is significantly related to exclusive breastfeeding practice. Mothers with a positive attitude toward exclusive breastfeeding are more likely to breastfeed exclusively than mothers with a negative attitude. According to Tadele, Habta, Akmel, and Deges (2016), mothers with positive attitude said they preferred to feed their children only with breast milk. They admitted that exclusive breastfeeding is better than artificial feeds.

Maternal breastfeeding self-efficacy and its association with breastfeeding outcomes have been highlighted in studies that promote breastfeeding. This study has discovered that the level of breastfeeding self-efficacy is significantly related to breastfeeding practice. Mother with high level of breastfeeding self efficacy were 8 times more likely to feed their baby until 6 month than those who were not high level of breastfeeding self-efficacy (Loke & Chan, 2013). According to Emmanuel (2015), strong commitment and self-efficacy increase the likelihood of breastfeeding for six months.

## 4 CONCLUSIONS

Maternal breastfeeding practice is governed by many variables. Maternal characteristics and breastfeeding self-efficacy are modifiable factors to which health professionals need to pay attention in order to improve the breastfeeding outcomes. Sociodemographic characteristics like age, education, knowledge, parity, occupation, and attitude towards breastfeeding may influence breastfeeding practice. Beside these factors, exclusive breastfeeding really depends on mothers' desire to breastfeed. When a mother is confident, exclusively breastfeeding her infant will succeed. Nurses, as the closest health professionals to mothers, need to assess these modifiable factors in order to enhance breastfeeding outcomes.

## REFERENCES


Bhanbhro, S. (2017). *Indonesia's Minangkabau culture promotes empowered Muslim women.* Retrieved from http://theconversation.com/indonesias-minangkabau-culture-promotes-empowered-muslim-women-68077.

Danso, J. (2014). Examining the practice of exclusive breastfeeding among professional working mothers in Kumasi Metropolis of Ghana. *International Journal of Nursing, 1*(1), 11–24.

Emmanuel, A. (2015). A Literature review of the factors that influence breastfeeding: An Application of the health belief model. *International Journal of Medicine and Health Science*, 2(3), 28–36.

Hafizan, N., Zainab, T. & Sutan, R. (2014). Socio-demographic factors associated with duration of exclusive breastfeeding practice among mothers in East Malaysia. *Journal of Nursing and Health Science, 3*(3), 52–56.

Loke, A.Y. & Chan, L. S. (2013). Maternal breastfeeding self-efficacy and the breastfeeding behaviours of newborn in the practice of exclusive breastfeeding. *Journal Obstetric Gynecology Neonatal, 42*(6), 672–684.

Maonga, A.R., Mahande, M.J., Mamian, D.J., & Msuya, S.E. (2016). Factors affecting exclusive breastfeeding among women in Muheza District Tanga Northeastern Tanzania: A mixed method community based study. *Maternal and Child Health Journal Springer, 20*(77), 77–87.

Ministry of Health Indonesia (2016). *Situation and analysis of exclusive of breastfeeding.* Retrieved from http://www.kemkes.go.id/development/site/layanan-kesehatan.

Ogunlesi, T.A. (2010). Maternal socio-demographic factors influencing the initiation and exclusivity of breastfeeding in a Nigerian Semi-Urban Setting. *Maternal and child health, 14(3), 459–465.*

Tadele, N., Habta, F., Akmel, D. & Deges, E. (2016). Knowledge, attitude and practice towards exclusive breastfeeding among lactating mothers in Mizan Aman Town Southwestern Ethiopia: descriptive cross-sectional study. *International Breastfeeding Journal, 11*(3), 1–7.

Ukegbu A.U, Ebenebe E.U. & Ukegbu PO. (2010). Breastfeeding pattern, anthropometry and health status of infants attending child welfare clinics of a teaching hospital in Nigeria. *South African Journal of Clinical Nutrition*, 23(4), 191–196.

World Health Organisation. (2015). Ten facts of breastfeeding. Retrieved from http://www.who.int/features/factfiles/breastfeeding/en.

Yati, R.M., Dienaputra, R.D. & Zakaria, M.M. (2014). The influence of the modernization of education toward Minang-girl's life in Sumatra's Westkust (1900–1942). *International Journal of Education and Research, 2*(6), 125–132.

*Strengthening Research Capacity and Disseminating New Findings in Nursing and Public Health – Malini et al. (Eds)*
*© 2018 Taylor & Francis Group, London, ISBN 978-1-138-50066-2*

# Depression and situational crisis on infertile Minangkabau women

Y.P. Sari & M.U. Ridwan
*Faculty of Nursing, University of Andalas, Padang, West Sumatra, Indonesia*

ABSTRACT: This study is aimed to explore more deeply the quality of life of Minangkabau women who are undergoing infertile therapy and the factors that influence it. This study used the qualitative approach to construct and find patterns of the quality of life of infertile women undergoing therapy. The total number of participants were seven people selected through a purposive sampling technique. The data was obtained using in-depth interviews from which precise transcripts were analysed using the "constant comparison" method. The study showed that Minangkabau women undergoing infertile therapy had a fluctuating stress level, prolonged sadness, and easily stimulated stress associated with the role and function of a wife. A long-term therapy process, at a high cost and with the uncertainty of the therapy results, also adds to psychological and financial burden for the couples. The belief in God's determination is significantly capable of creating a positive self-concept on women with infertile problems.

## 1 INTRODUCTION

Infertility is an unpleasant experience and tends to cause distress and depression in women (Xiaoli et al., 2016; Namdar et al., 2017). The community's stigmatisation of infertile women as being "barren women" often becomes one of the factors that exacerbates the sense of shame and sadness that they feel (Kaadaaga et al., 2014)

One study Tabong & Adongo (2013) found that many women with infertility envy other women who have children, feel angry when they see children scolded by their biological mother, and sometimes blame themselves for the inability to have a child. Infertile women often experience a problematic relationship with their partners, such as lack of sex, the risk of physical violence from the husband, and being neglected, which could even lead to divorce. Although another study found that both women and men experienced self-concept disorder, self-acceptance, and emotional stress associated with this infertility problem (Haica, 2013), women are at risk of suffering from more severe disorders.

For Minangkabau women, the ability to have children is somewhat "mandatory" in the eyes of the community (Bennett et al., 2012). Consequently, this cultural expectation becomes the main reason that causes infertile women to perform a series of therapies. Different types of therapy ranging from simple to complex forms are time-consuming and expensive. This becomes another form of stress that can exacerbate the women's psychological problems because, on one hand, they have to do a variety of exhausting therapy. Sometimes they feel ashamed when being subjected to genital examinations and they must also consume a lot of medicine. On the other hand, the results of treatment cannot be predicted from the beginning. Minangkabau women are considerably burdened with infertile status because family lineage will be cut off if they cannot give birth to an offspring.

## 2 METHODS

This study is aimed to identify factors that affect the quality of Minangkabau women who are undergoing both medical and alternative infertile therapy. Thus, the approach used is

Grounded Theory (Corbin & Strauss, 1990). Sampling technique used in this study is theoretical sampling involving seven Minangkabau women who are undergoing infertile therapy and have been married between 4–10 years. Data were collected through in-depth interviews and field observations by means of tape recorders and field notes for further verbatim transcripts. The interviews were conducted in Indonesian and Minangkabau language and later translated into Indonesian and English. Data were analysed using "constant comparison" technique, i.e. data processing method to analyse concept systematically, categorise, and interpret data from research question. Research ethics in this research is obtained from Medical and Health Research Ethics Committee, Faculty of Medicine Andalas University.

## 3 RESULTS

Participants of this study comprise seven Minangkabau women who were undergoing medical therapy in several hospitals and alternative therapy. In this session, situational depression and crisis, which are the "core" category of research that explains the feelings of Minangkabau women undergoing infertile therapy, will be discussed.

### 3.1 *Participant characteristics*

Participants in this study are women with an age range of between 25 and 40 years, and a duration of marriage of between two and ten years. All participants have a gap of two to four years between them and their husbands. The level of education varies from senior high school graduates to Master's degree.

### 3.2 *Depression and situational crisis*

This category illustrates that Minangkabau women who are undergoing therapy feel that they are in an uncomfortable situation both psychologically and physically. This is evident from some expressions of the following participants:

> Really, being in this position makes me feel uncomfortable. I really feel very depressed and sometimes make me try to limit the interaction with other family members ...
>
> *(P1)*
>
> It is not easy to be in my position. When everyone demands me to be able to give birth to the family without thinking how hard my efforts have been in this therapeutic process, I feel intimidated...
>
> *(P3)*

Another participant reveals that at certain moments, she can forget the sadness she experienced but the feeling will reappear when she is in solitude. As the participant expresses:

> Indeed, I do not always feel sad all the time. I keep trying to be strong. However, almost every night when I was alone, especially after prayer, I cried to express disappointment. I felt alone. No one else really understands what I feel ...
>
> *(P5)*

Physical discomfort is also expressed by participants when undergoing a series of therapies, as expressed by the following participants:

> I have undergone this therapy for some time. There has been a transvaginal examination. Then, I feel very shy and painful. Especially when the HSG examination, it feels very painful...
>
> *(P6)*
>
> I was given a hormone therapy by a doctor but the effect that I felt made me feel like I want to stop doing this therapy, instead of getting pregnant. I was getting fatter and acne was on my face...
>
> *(P4)*

### 3.3 *Low self-esteem*

Participants in this study feel that they are not perfect because they are unable to give offspring to their families. This often becomes a psychological burden for them. Some participants revealed:

> Every time I see other people get pregnant, even my co-workers, it makes me sad. I asked myself: what am I lack of? Why God did not trust me to be a mother, like any other woman…
>
> *(P7)*
>
> Whenever I am invited to gather with my co-workers, I try to avoid it. I feel alienated because they tell me more about their children. It made me feel different…
>
> *(P6)*
>
> There was a neighbour who made me feel very demeaned when she said: What is going on with you? Almost eight years married but aren't pregnant yet. You can later be left by husband. No one wants to get married without having children afterward. And I cried for myself after that…
>
> *(P4)*

### 3.4 *Fatigue*

Some participants who are undergoing infertile therapy feel tired of the long and expensive therapy situation and process that have unpredictable results. Some participants revealed:

> Initially, my husband accompanied me doing this therapy; each consultation session. But on the final inspection, he left me alone for some business reasons. At that time, I thought, it seems only me who tried to get pregnant myself, really. To be honest, I'm tired, I'm tired...
>
> *(P4)*
>
> Long queues, tense situations when it comes to therapy. It makes me think it is just wasting time. I'm tired, when I have to run for it amidst my busy work…
>
> *(P2)*
>
> After six months of treatment, I feel tired because I consume a variety of supplements everyday by which I cannot feel the results until now…
>
> *(P1)*

### 3.5 *Negative self-perception*

Many participants see themselves negatively, thinking that what they are experiencing today is a punishment for their past mistakes and past husband mistakes. They feel that they do not deserve to be mothers. This is indicated by some of the participants' expressions. Participant 3 said that she really feels like being a 'failed' woman and wife and wonders if she can be happy without having a child. Another participant also said:

> I feel, I am being punished by God for the wrongs that my husband and I might have done. I try to remember, what is the real mistake that made God punish me like this? Punish me because I might be considered as an inappropriate mother to be…
>
> *(P5)*

## 4 DISCUSSION

Minangkabau women undergoing infertile therapy feel depressed and are in situational crisis throughout the course of therapy. This happens as a manifestation of their failure to carry out roles and functions as women and wives where it is difficult for them to accept the inability to have offspring (Karaca and Unsal, 2015).

The infertile women express such physiological responses as blaming themselves, feeling lonely, and feeling imperfect and embarrassed. Several earlier quantitative studies found that depression and anxiety were the main psychological responses in infertile women with a prevalence of 44.8% (Lakatos et al., 2017). This is exacerbated when the prevailing cultures require women to be able to produce offspring / a future generation. Minangkabau society has such cultural demands. The same issue has been studied in Ghana where a strong culture with "the obligation to give birth" has been a major source of conflict for couples, causing women to isolate themselves from their social environment as compensation for depression (Alhassan et al., 2014). To live in such a demanding culture adds to the psychological burden of women with infertility so that most of them try various therapies to overcome the problem. This is confirmed by a survey in Indonesia that shows that in recent years there has been an increase in women's visits to places providing infertility programme services and therapies, 92% of which attended OBSGYN doctors (Bennett et al., 2012).

Various kinds of infertile female responses arise when undergoing this therapy. Their responses include the low level of satisfaction with the quality and process of the therapy services, lack of acceptance of final verdicts, high costs, types of examinations to be followed, poor service, and increased risk of depression in partners, especially in women (Sexty et al., 2016; Lakatos et al., 2017). Examinations of reproductive organs often cause women to feel ashamed. Some women often complain about the procedure that requires them to expose their intimate sex organs in front of a male doctor. Besides, some types of examination sometimes cause pain in the females' reproductive organs, such as USG transvaginal examination and Hysterosalpingography (HSG).

Such discomfort felt by Minangkabau women who are undergoing infertile therapy is also due to pressure from their extended families. This pressure usually comes from the female's family in response to her sterility. Some participants in the study confessed that they actually felt stress when facing their own big family. The extended family members constantly ask about many things in connection with the business of having offspring (becoming pregnant). This occurs as a cultural consequence for those born and raised in Minangkabau where the family lineage originates from the mother (matrilineal system).

In contrast to patrilineal culture where lineages originate from the father's side, women are much more likely to get pressure from their husbands because they feel rejected in their social systems (Taghavi et al., 2015). These pressures often make women feel intimidated. In line with the above studies, participants in this study felt intimidated by questions from family, friends or neighbours about their infertility status. Some women felt disturbed and cornered by such inquiries so that they tried to isolate themselves from their surroundings. The pressures infertile women undergo become an irony because a study suggests that social support from the family significantly affects the infertile women's quality of life (Fu et al., 2016).

Most participants have low self-esteem. Some participants have claimed that they are embarrassed when they have to gather in social activities. On these occasions other people come by bringing their entire family members, whilst they are only accompanied by a spouse. More often, a joke made by friends and neighbours makes them feel inferior, owing to the stigma of being barren (Namdar et al., 2017).

Moreover, a participant admitted to feel that what she experienced was a form of punishment God has given her for the sins and mistakes of herself and her husband in the past. They feel that God assumes that they do not deserve to be parents, even if they claim to have everlasting love if one day they were to have children.

A quantitative study by Rashidi et al., (2008) found low self-esteem that affects the quality of life of infertile women is more common in women with low levels of education and in younger women. The feeling of inferiority arises because they feel imperfect as a woman because of their inability to continue the family lineage.

Some literature confirms that there is a negative impact of depressive and crisis feelings experienced by women with infertility with their ovarian function, which affects the success of therapy. However, a longitudinal study conducted on 217 heterogeneous couples found no effect of stress, anxiety, or depression experienced by infertile women with physiological ovaries (Donarelli et al., 2016). Nevertheless, psychosocial problems hamper relationships between partners and sometimes affect family harmony.

## 5 CONCLUSION

Quality of life is a person's expectation of life compared to the reality faced. WHO establishes six measures of quality of life, namely physical, psychological, social, spiritual, religious and environmental domains. Minangkabau women who experience infertility undergo disruptions in almost all domains. These disturbances in the domains of quality of life are manifested in the emerging psychological disorders, such as depression, being trapped in long-term crisis situations, having less supportive environments, and isolation from social relationships. These disturbances need serious attention from health workers, especially nurses. It is necessary to develop a nursing intervention that focuses on efforts to suppress psychological disorders in infertile women, especially when they undergo the therapeutic process. Thus, their quality of life is expected to improve and be satisfactory.

## REFERENCES


Alhassan, A., Ziblim, A.R., & Muntaka, S. (2014). A survey on depression among infertile women in Ghana. *BMC Womens Health, 14*(1), 42, 1–6.

Bennett, L.R., Wiweko, B., Hinting, A., Adnyana, I.B.P. & Pangestu, M. (2012). Indonesian infertility patients' health seeking behaviour and patterns of access to biomedical infertility care: An interviewer administered survey conducted in three clinics, *Reproductive Health, 9*, 24, 1–7.

Corbin, J. & Strauss, A. (1990). Grounded theory research: Procedures, canons, and evaluative criteria. *Qualitative Sociology, 13*(1), 3–21.

Donarelli, Z., Lo, G., Gullo, S., Marino, A., Volpes, A., Salerno, L. & Allegra, A. (2016). Infertility-related stress, anxiety and ovarian stimulation: Can couples be reassured about the effects of psychological factors on biological responses to assisted reproductive technology? *Reproductive Biomedicine Society Online, 3*, 16–23.

Fu, B., Yan, P., Yin, H., Zhu, S., Liu, Q., Liu, Y. & Lei, J. (2016). Science direct psychometric properties of the Chinese version of the infertility self-efficacy scale. *International Journal of Nursing Sciences 3*(3), 259–267.

Haica, C.C. (2013). Gender differences in quality of life, intensity of dysfunctional attitudes, unconditional self-acceptance, emotional distress and dyadic adjustment of infertile couples. *Procedia—Social and Behavioral Sciences, 78*, 506–510.

Kaadaaga, H.F., Ajeani, J., Ononge, S., Alele, P.E., Nakasujja, N. & Manabe, Y.C. (2014). Prevalence and factors associated with use of herbal medicine among women attending an infertility clinic in Uganda. *BMC Complementary and Alternative Medicine, 14*(27), 6–8.

Karaca, A. & Unsal, G. (2015). Psychosocial problems and coping strategies among Turkish women with Infertility. *Asian Nursing Research, 9*(3), 243–250.

Lakatos, E., Szigeti, J.F., Ujma, P.P., Sexty, R. & Balog, P. (2017). Anxiety and depression among infertile women: A cross-sectional survey from Hungary. *BMC Womens Health 17*(48), 1–9.

Namdar, A., Naghizadeh, M.M., Zamani, M. & Yaghmaei, F. (2017). Quality of life and general health of infertile women. *Health and Quality of Life Outcomes, 15*(139), 1–7.

Rashidi, B., Montazeri, A., Ramezanzadeh, F., Shariat, M., Abedinia, N. & Ashrafi, M. (2008). Health-related quality of life in infertile couples receiving IVF or ICSI treatment. *BMC Health Services Research, 8*(186), 1–6.

Sexty, R.E., Hamadneh, J., Rösner, S., Strowitzki, T., Ditzen, B., Toth, B. & Wischmann, T. (2016). Cross-cultural comparison of fertility specific quality of life in German, Hungarian and Jordanian couples attending a fertility center. *Health and Quality of Life Outcomes, 14*(27), 1–8.

Tabong, P.T. & Adongo, P.B. (2013). Infertility and childlessness: A qualitative study of the experiences of infertile couples in Northern Ghana. *BMC Pregnancy and Childbirth, 13*(72).1–7

Taghavi, S.A., Bazarganipour, F., Hugh-Jones, S. & Hosseini, N. (2015). Health-related quality of life in Iranian women with polycystic ovary syndrome: A qualitative study. *BMC Women's Health, 15*(111), 1–8.

Xiaoli, S., Mei, L., Junjun, B., Shu, D., Zhaolian, W., Jin, W. & Ju, Q. (2016). Assessing the quality of life of infertile Chinese women : A cross-sectional study. *Taiwanese Journal of Obstetrics & Gynecology 55*(2), 244–250.

# *Mental health*

*Strengthening Research Capacity and Disseminating New Findings in Nursing and Public Health – Malini et al. (Eds)*
*© 2018 Taylor & Francis Group, London, ISBN 978-1-138-50066-2*

# Coping mechanisms and stress levels in thesis writing among nursing students

F. Fernandes & Dahlia
*Faculty of Nursing, Andalas University, Padang, West Sumatra, Indonesia*

Basmanelly
*HBSaanin Mental Hospital, Padang, West Sumatra, Indonesia*

ABSTRACT: The aim of this research is to identify the relationship between coping mechanisms and stress levels among nursing students at Andalas University. The sample for this research constituted 124 respondents, chosen through survey (total sampling) method. Data was collected at the Faculty of Nursing, Andalas University during the period 15–23 May 2017. This study shows that almost half (45.2%) of the students had high levels of stress and more than half (54.8%) of the respondents used maladaptive coping mechanisms. There was also a significant relationship ($p = 0.000$) between coping mechanism and stress level among the students when completing their theses. The result of this study also indicates that nursing students should seek emotional support from others to overcome the problems that arise during their thesis completion, so that their stress level maybe reduced.

## 1 INTRODUCTION

Mental health is very important for human beings. Mental health should be a healthy emotional, psychological, and sociological condition, seen from satisfactory interpersonal relationships, behaviour and effective coping, positive self-concept, and emotional stability (Videbeck, 2011). According to Stuart and Laraia (2012), the state of mental health is associated with happiness, satisfaction, accomplishment, and hope.

According to the World Health Organization (2017), around 35 million people experienced stress in 2016, with 60 million people affected by bipolar disorder, 21 million people affected by schizophrenia, and 47.5 million people were exposed to the dementia. Data from *Riskesdas* (2013) shows that the prevalence of mental disorders, the emotional symptoms of depression, and anxiety have over the past 15 years affected around 14 million people, or 6% of the Indonesian total population, while the prevalence of severe mental disorder, like schizophrenia, has affected 1.7% of inhabitants. The prevalence of mental disorder in West Sumatra was 4.5%,with 1.9% experiencing severe psychiatric disorders, which means that the province has the ninth highest number of cases of mental disorder of the 33 provinces in Indonesia (Riskesdas, 2013).

Stress can occur at different age levels and in different employments, including that of students. Results from the National College Health Assessment's (2013) research into 125,000 students from 150 colleges and universities in the United States revealed that 30% of the students experienced stress (American Psychological Association, 2014). Another investigation of 1,224 students in India showed that 299 students (24.4%) experienced stress, with 10% having a heavy level of stress prevalence, 7.6% a moderate level of stress, and 6.8% a low level of stress (Waghachavera, 2013).

Stress is caused by the presence of a stressor. A stressor is a situational stimulus that reduces one's ability to feel happy, comfortable, and productive (Saam & Wahyuni, 2012). Research conducted on 1,400 students of Midwestern University showed that stressors on students can originate from academic demands and financial pressures (40%), problems with friends or other social relations (27%), career concerns (22%), and physical appearance (20%) (Prichard, 2012). This result is in line with research conducted by Eva et al. (2015) that demands for academic excellence, one of which is thesis completion, is the main stressor for college students.

Failure in thesis completion can result from the difficulty in finding literature and reading materials, bad coordination between students and thesis adviser, and lack of ability to do the research (Broto, 2016). When these problems generate pressure on the students, stress in completing their theses can manifest itself (Gunawati et al., 2006). Scarfi's (2014) study of 374 students at Andalas University who were completing their theses indicated that 72.2% of the students experienced stress, 15.2% experienced moderate stress, and 12.6% experienced severe stress.

Stress overload (distress) can be a threat to a student's academic performance. One example is academic procrastination, which is a kind of delay made to completing formal types of assignments related to academic tasks (Ferrari et al., 1995). This is in line with the results of Andarini's (2013) research finding that the higher the distress, the more that academic procrastination occurs, and vice versa. Yanti's (2016) investigation of teenagers at MTsN (Islamic Junior High School) LubukBasung noted that 81 of the respondents (52.3%) reported high academic procrastination and 74 of respondents (47.7%) dealt with low academic procrastination.

In a preliminary study on 18 April 2017, ten students of Programme A of the Nursing Faculty who were writing a thesis were interviewed. The study found that four people felt nervous and depressed when dealing with their thesis. Two students said that they tended to eat excessively, and two others said they tried to get help from others. Four people admitted that they had become more easily upset because a lot of things were going on that were out of their control. Two people said they preferred to extend their sleeping time to control emotions, one student preferred watching TV to forget problems, and another performed more prayer. Further, one student admittedthat she did not feel capable of coping with the problems and blamed herself for what happened to her. Another person said that s/he was able to overcome various problems and disorders that occur every day by trying to look at current issues with a positive outlook. Based on these phenomena, the researchers are interested in investigating 'the relationship between coping mechanisms and stress levels among nursing students of Andalas University preparing theses'.

## 2 METHOD

This is descriptive analytic research using a cross-sectional approach in order to identify the relationship between coping mechanisms and stress levels among nursing students of Andalas University when preparing their theses. Research was carried out between February and July 2017. The sample was 135 students of Programme A (2013) selected by using total sampling technique.

The data-collecting instruments used in this study consisted of the Perceived Stress Scale-14 (PSS-14) questionnaire to measure stress levels in nursing students and the Brief COPE Inventory (BCI). Data was analysed using univariate and bivariate analysis. Univariate analysis aims to identify the distribution of respondent characteristics, coping mechanisms, and stress levels. Bivariate analysis is applied to recognise the relationship between coping mechanisms and stress levels. A chi-squared test was used for the bivariate analysis.

## 3 RESULTS

### 3.1 *Description of coping mechanism and stress level*

Table 1 reveals that 68 of the respondents (54.8%) used the maladaptive coping mechanism.

Table 2 shows that 56 of the respondents (45.2%) experienced heavy stress levels.

Table 1. Distribution of coping mechanisms.

| Coping mechanism | *f* | % |
|---|---|---|
| Adaptive | 56 | 45.2 |
| Maladaptive | 68 | 54.8 |
| Total | 124 | 100 |

Table 2. Distribution of stress levels.

| Stress level | *f* | % |
|---|---|---|
| Low | 15 | 12.1 |
| Moderate | 53 | 42.7 |
| Heavy | 56 | 45.2 |
| Total | 124 | 100 |

Table 3. Relationship between coping mechanism and stress level.

| Coping mechanism | Stress level (%) Low | Moderate | Heavy | Total (%) | *p* value |
|---|---|---|---|---|---|
| Adaptive | 11.3 | 25.8 | 8.1 | 45.2 | 0.000 |
| Maladaptive | 0.8 | 16.9 | 37.1 | 54.8 | |
| Total | 12.1 | 42.7 | 82.3 | 100 | |

### 3.2 *Relationship between coping mechanism and stress level*

The chi-squared test results shown in Table 3 indicate a *p* value of 0.000, which confirms a meaningful relationship between the coping mechanisms and the students' stress levels in preparing theses in the Nursing Faculty at Andalas University.

## 4 DISCUSSION

The results of the research on the coping mechanisms of students in completing theses at the Nursing Faculty of Andalas University, Padang showed that more than half of the students (54.8%) use maladaptive coping mechanisms. This result supports Puspitasari's (2014) study on the nursing undergraduate students of the 8th semester of *FikkesUnimus* (Faculty of Nursing and Health, University of Muhammadiyah, Semarang),which showed that more than half of the students (58.9%) used maladaptive coping mechanisms.

Anadaptive coping mechanism is a form of mechanism that supports the integration of the functions of coping, growth, learning, and achieving goals. The category includes talking with others, solving problems effectively, performing relaxation techniques, and exercising balanced and constructive activity. Meanwhile, amaladaptive mechanism is a form of coping that inhibits the functions of integration and growth, reduce autonomy and tends to dominate the environtment. The category is manifested through overreacting, unwillingness to eat, work overload, and avoidance(Stuart & Laraia, 2012).

Coping mechanisms area way in which individuals resolvea problem, adjust to change, and respond to threatening situations (Keliat & Akernat, 2011). However, everyone has a different approach to tackling and overcoming stress. In general, the coping happens automatically when individuals feel the existence of stressful situations or threats. The individuals are then required to overcome the tension that they experience as soon as possible. Afterwards, individuals mayconductan evaluation to decide the coping mechanism best applied. Coping reactions towards problems vary from one individual to another, and from time to time in the same individual (Smeltzer & Bare, 2013).

Our research shows that nearly half of the nursing students (45.2%) experienced heavy levels of stress in completing their theses. This study discovered that the majority of the respondents (83%) feel nervous and depressed when facing problems in writing their theses. The students list the heavy demands from the environment as the cause, which they consider as threats. In these cases, they would feel the pressure and then finally the stress. This result supports the findings of a previous study (Syofia, 2015) in another faculty of nursing, which highlighted moderate and heavy stress levels (88.9%) among students. Sarafino and Smith (2011) claim that stress occurs due to inaccurate perceptions in relation to environmental demands and available resources. Everyone will experience different pressures from the same stressor.

We assume that the stress experienced by the majority of our respondents is also influenced by gender factors, because most respondents in this study were women (96%). This is confirmed by the results of the research in the United States, claiming that women tend to have higher levels of stress than men. In general, women experience stress 30% more than do men (Gunawatiet al., 2006). In 2010, the picture of stress in America was illustrated by the American Psychological Association (APA, 2010). It reported statistics about women and stress, saying that 49% of women experience high levels of stress. Based on the theory above, the conclusion can be drawn that women have higher levels of stress than men.

Our study also shows that the heaviest level of stress occurs in8.1% of those students using adaptive coping mechanisms, and in 37.1% of those using maladaptive coping mechanisms. A chi-squared test obtained a*p* value of 0.000, confirming that there is a meaningful relationship between coping mechanism and stress level among the nursing students of Andalas University. This finding supports the same conclusions of previous studies such as those of Dwipermana (2016) at Ngudi Waluyo School of Health, and Wijayanti (2013) at the Faculty of Education, State University of Yogyakarta.

In preparing their theses, the students are confronted by a variety of both internal and external stressors. If their chosen coping mechanisms are incapable of dealing with a stressor, the stress level will continue to grow greater and severe depression may even manifest and trigger self-destructive actions such as suicide. This is in accordance with Pheukphan's (2009) argument that the heavy level of stress among nursing students will provide a very significant and visible impact, characterised by such symptoms as depression and even the possibility of suicide.

In addition, emotional and social support from friends and parents need consideration in order to have an appropriate and effective coping mechanism. This can be inferred from thecurrent research in which of all the respondents experiencing a moderate level of stress (15 people), more than half (53.3%) admitted that they often asked for social and emotional support from others, and 40% of the respondents reported doing the same thing. This is in accordance with what Boyd (2012) claimed: that social support is very important in helping people to cope with stress. Success in finding a coping mechanism to deal with stress improves quality of life, and physical and mental health.

This study suggests that students who are writing a thesis should apply an adaptive coping mechanism, which is evidently effective in resisting the stressor being faced, by accepting the problem, thinking positively, trying to find sources of support from people nearby, and seeking religious solace. Students will then be able to better cope with the stress caused by their thesis—related problems. In the meantime, students who apply a maladaptive coping mechanism through refusing to cope with the facts and thinking negatively will usually be unable to cope with the stress-triggering problems and may even experience higher levels of stress.

## 5 CONCLUSION

The study has identified a relationship between coping mechanism and stress level. More than half of the students used maladaptive coping mechanisms and almost half of the students experienced severe levels of stress in preparing their theses. We recommend that students find emotional support from others to overcome the problems that arise during the completion of their theses, which can reduce their stress levels. As to the Faculty of Nursing, the results of this study provide information that is expected to be used as a consideration when making policies for final-year students, guiding them to prepare their theses before the pre-clinic in the 8th semester, so that the focus of students is not divided and their stress is reduced when preparing their theses. We recommend that further research should examine more details about forms of action that can be taken to reduce the incidence of stress and about ways to manage stress with appropriate coping mechanisms among nursing students.

## REFERENCES


American Psychological Association (APA). (2010). *Gender and stress*. Washington, DC: American Psychological Association. Retrieved from http://www.apa.org/news/press/releases/stress/2010/gender-stress.aspx.

American Psychological Association (APA). (2014). *Student Under Pressure*. Washington, DC: American Psychological Association. Retrieved from http://www.apa.org/monitor/2014/09/cover-pressure.aspx.

Andarini, S.R. (2013). Distress and Social Support With Academic Procrastination. *Jurnal Talenta Psikologi Vol. 2(2)*.

Arikunto, S. (2010). *Research Procedure a Practical Approach*. Jakarta, Indonesia: RinekaCipta.

Banarjee, N.& Chatterjee, I. (2016). Academic stress, suicidal ideation & mental well-being among 1st semester & 3rd semester medical, engineering & general stream students. *Journal of Art, Science and Commerce*, *7*(3), 73–80.

Boyd, A. (2012). *Psychiatric nursing: Contemporary practice* (5thed.). Philadelphia, PA: Wolters Kluwer Health. Broto, H. (2016). *Stress in Thesis Writing Among Students* (Unpublished undergraduate thesis, Sanata Drama University, Yogyakarta, Indonesia).

Dewanti, D.E. (2016). *Tingkat stres akademik pada mahasiswa bidik misi dan non bidikmisi Fakultas Ilmu Pendidikan* (Unpublished undergraduate thesis, Universitas Negeri Yogyakarta, Yogyakarta, Indonesia).

Dwipermana. (2016). *Coping Mechanism and Stress in Thesis Writing Among Students* (Unpublished undergraduate thesis, Ngudi Waluyo School of Health, Semarang, Indonesia).

Eva EO, Zakirul Islam, Abu SM, Faizur Rahman, Rini JR, Hassan Iftekhar et al. (2015). Prevalence of stress among medical students: a comparative study between public and private medical schools in Bangladesh. Retrieved from http://bmcresnotes.biomedcentral.com/articles/10.1186/s13104-015-1295-5.

Ferrari, J.R., Johnson, J.L., & McCown, W. (1995). *Procrastination and task avoidance: theory, research, and treatment*. New York: Plenun Press.

Gunawati, R., Hartati, S. & Listiara, A. (2006). *Student and Supervisors' Effectiveness of Communication and Stress in Thesis Writing. Jurnal Psikologi Universitas Diponegoro*, *3*(2), 93–115.

Keliat, B.A. & Akermat, S. (2011). *Community Mental Health Nursing*. Jakarta, Indonesia: EGC.

Pheukphan, A.P. (2009). *Stressors and coping strategies among AU nursing student*. Bangkok, Thailand: Faculty of Nursing Science, Assumption University of Thailand. Retrieved from http://www.nurse.au.edu/index.php/academic-activity/83-stressors-and-coping-strategies-among-au-nursing-student.html.

Potter, P.A. & Perry, A.G. (2013). *Fundamentals of nursing*. Philadelphia, PA: Elsevier.

Prichard, J.R. (2012). Academic Stress, Social Trauma, and Disturbed Sleep in A Large Population of College Students: Interconnections and Health Implications. *Journal of Adolescent Health. Vol. 50(2) S32*.

Puspitasari, I. (2014). *Coping Mechanism in Thesis Writing Among Nursing Students* (Unpublished undergraduate thesis,University ofMuhammadiyah, Semarang, Indonesia).

Riskesdas. (2013). *Laporan Hasil Riset Kesehatan Dasar (Riskesdas) 2013*. Jakarta, Indonesia: Kementrian Kesehatan RI.

Saam, Z., & Wahyuni, S. (2012). *Nursing Psychology*. Jakarta: Rajawali Pers.

Sarafino, E.P. & Smith, T.W. (2011). *Health psychology: Biopsychosocial interactions* (7th ed.). New York, NY: John Wiley & Sons.

Scarfi, F. (2014). *Self Efficacy,* Social Support and Stress Levels in Writing Thesis Among Students. (Unpublished undergraduate thesis). Andalas University, Padang, Indonesia.

Smeltzer, S.C. & Bare, B.G. (2013). *Brunner and Suddarth's Texbook of Medical Surgical Nursing* (13th ed.). Jakarta, Indonesia: EGC.

Stuart, G.W. & Laraia, M.T. (2012). *Principles and practice of psychiatric nursing* (8th ed.). St. Louis, MO: Mosby.
Syofia, E. (2015). *Causes of Student's Stress.* (Unpublished undergraduate thesis. University of Northern Sumatra, Indonesia).
Videbeck, Sheila L. (2011). *Psychiatric-Mental Health Nursing (5th ed).* Philadelpia: Wolters Kluwel Health.
Waghachavera, Vivek B. (2013). *A Study of Stress among Students of Professional Colleges from an Urban area in India.* Retrieved from https://www.ncbi.nlm.nih.gov/pmc/articles/pmc3749028/.
Wijayanti, N. (2013). *Coping Mechanism in Thesis Writing Among Nursing Students.* (Unpublished undergraduate thesis. Universitas Negeri Yogyakarta.
World Health Organization. (2017). *Global health observatory data.* Geneva, Switzerland: World Health Organization. Retrieved from http://www.who.int/gho/mental_health/en/.
Yanti, F.N. (2016). *Parenting Style and Academic Procrastination at MTsN 1 LubukBasung*(Unpublished undergraduate thesis, Andalas University, Padang, Indonesia).

*Strengthening Research Capacity and Disseminating New Findings in Nursing and Public Health – Malini et al. (Eds)*
*© 2018 Taylor & Francis Group, London, ISBN 978-1-138-50066-2*

# Relationship between coping mechanism and stress level among parents with mentally retarded children at SLB Wacana Asih, Padang

I. Erwina, D. Novrianda & W. Risa
*Faculty of Nursing, Andalas University, West Sumatra, Indonesia*

ABSTRACT: This research attempts to identify the relationship between coping mechanisms and stress levels among parents with mentally retarded children at a school for disabled children (SLB Wacana Asih) in Padang, Indonesia. Method: The sample was 55 respondents gathered through a census (total sampling) method. Result: 85.5% of the respondents used an adaptive coping mechanism and 54.5% of respondents had a high stress level. There was a correlation ($p = 0.006$) between coping mechanism and stress level among parents with mentally retarded children at SLB Wacana Asih, Padang. Conclusion: The study suggests that parents are recommended to use an adaptive coping mechanism in taking care of their mentally retarded children. It is suggested that health workers conduct health promotion about adaptive coping mechanisms and how to cope with stress for parents with mentally retarded children.

*Keywords*: coping mechanism, mentally retarded children, parents, stress

## 1 INTRODUCTION

Basically, parents expect children to have a normal physical, psychological, and cognitive development. Most parents find it difficult to accept the reality of having a child born in a state of imperfection or with a development defect, such as mental retardation (Mangunsong, 2012). Mental retardation is a condition in which a person's intelligence is well below average and is characterised by limited intelligence and incompetence in social communication. This condition is commonly known as mental retardation due to the limitation of intelligence, which leads to difficulty in following ordinary school education (Kosasih, 2012).

Based on World Health Organization (WHO) statistics (Ministry of Health, 2014), an estimated 11,8% of the population in developed countries and 18% in developing countries experience disability, where the population of children with mental retardation is much higher if compared to the number of children with other limitations. Of the 220 million population in Indonesia, 6.6 million are people with mental retardation (Sen & Yurtsever, 2007, in Nataya, 2014).

Ministry of Education and Culture (2015) stated that based on the summary of *Sekolah Luar Biasa* (SLB) [School for Disabled Children], there are 131 SLBs in West Sumatra. Padang Education Office stated that for the academic year 2015, there were 37 SLBs in Padang with a total of 1,331students. The number of students with mental retardation is higher in the population if compared to other types of disability, that is, cumulatively 735 students. Of the 37 SLBs in Padang, SLB Wacana Asih has the highest number of students (91 pupils) and 55 of the are children with mental retardation. A mentally retarded child cannot be self-sufficient as an individual in performing his or her own daily activities and has a limited ability to understand social behaviour and develop social skills (Wenar & Kerig, 2007). This mentally retarded child needs the closest person to help him/her in matters s/he cannot deal with on their own. Family is the first social environment for the child that will greatly influence his/her development. Parents are also regarded as the leading mentors for this type of child in later life because parents are expected to be heavily involved, and to participate in every part of the education and training of their children (Dardas and Ahmad, 2015).

The burden and pressure of taking care of a child with mental retardation appears to have an impact and influence on parents' psychological function, behaviour, and reactions. From an interactive perspective, it appears that stress experienced by parents comes from their inability to adapt to their child's daily care demands. The fact that parenting mentally retarded children takes an uncertain time and presents an unpromising future becomes the reason why parents lose control of the situation and why they suffer from severe stress (Karasavvidis, 2011). Some researchers have examined the parameters that have an impact on the creation of stress. It can be concluded that factors that play an important role cover the diagnosis of the child's disability, the seriousness of the child's disability, the child's behaviour, the source of family support, and the quality of the husband–wife relationship Karasavvidis, 2011). Meanwhile, according to (Bauman (2004), influential and determining factors of stress in caring include coping skills, problem-solving, religiosity, child attitudes, marital status and satisfaction, education and employment of parents, and the children's health status.

Research conducted by Hidangmayum and Khadi (2012) in India showed that parents of children with mental problems had higher stress levels (73.4%) than parents with normal children (21.7%). Another study by Islam et al. (2013) in Bangladesh also indicated significant differences between the stress experienced by parents raising mentally retarded children and those parents who had normal children, with parents with mentally retarded children having a higher level of stress.

The effects of unrevealed stress can lead to physical and mental illness, which can ultimately decrease productivity and cause poor interpersonal relations (Rasmun, 2004). The stress level experienced by an individual may go up and down, depending on how the individual adapts to the source of the stress. This process of adaptation is called a coping mechanism (Lazarus & Folkman, 1984; Davidson et al., 2006, in Oktavira, 2015).

A coping mechanism creates two goals. First, the individual tries to change the relationship between him/herself and his/her environment in order to produce a better effect. Secondly, individuals usually seek to relieve or eliminate the burden of coping and emotional stress (Safaria, 2009). A coping mechanism is an effort to defend a sense of control over situations that reduce comfort and to deal with stressful situations (Videbeck, 2008). Coping mechanisms are divided into two categories: adaptive and maladaptive. Adaptive coping mechanisms are coping that supports the function of integration, growth, learning, and achieving goals, whereas maladaptive coping mechanisms are coping that inhibits integration, breaks growth, lowers autonomy, and tends to dominate the environment (Stuart, 2013).

A study conducted by Suri and Daulay (2012) on parents with mentally retarded children in the Deli district of North Sumatra showed that 98.4% of the parents had applied adaptive coping mechanisms. Another study (Swari, 2014) on parents who had mentally retarded children in Denpasar also showed that most parents (98.75%) used an adaptive coping mechanism.

## 2 METHOD

### 2.1 *Study design*

This is a descriptive analytical and cross-sectional study. The population of this study is all parents of children with mental retardation in SLB Wacana Asih, Padang.

### 2.2 *Sample and sampling technique*

The sampling technique used is a census (total sampling) with the number of samples to be studied being as many as 55 respondents. Parents who became respondents in this study had met the following criteria: a) parents (either father or mother) of a child with mental retardation studying at SLB Wacana Asih, Padang; b) willing to participate as respondents; c) cooperative and literate.

### 2.3 *Ethical consideration*

Ethics approval was acquired from the Faculty of Medicine, Andalas University, Padang, Indonesia. Ethical approval was obtained by sending the completed ethical form to the Ethics Committee. After getting ethical approval, the next step was to recruit participants.

### 2.4 *Data collection and data analysis*

Data was collected by using two instruments for measuring coping and stress level. Brief COPE (Carver, 1997) was applied to measure coping mechanism, and Parenting Stress Index-Short Form (PSI-SF) was used to measure stress level. The researcher also recorded the participants' demographic characteristics. The collected data was then processed using univariate and bivariate analysis. Univariate analysis was applied to produce frequencies and percentages for the data, while bivariate analysis (chi-squared test through IBM SPSS statistical software) was used to measure the relationship between coping and stress.

## 3 RESULTS

The demographic characteristics showed that most respondents (34 people or 61.8%) were in the 31–40 age group. Females formed the majority of respondents (as many as 50 people or 90.9%) and more than half of the respondents (30 people or 54.5%) had a high school education. In 2016, most mentally retarded children (32 students or 58.2%) in SLB Wacana Asih, Padang were male and as many as 30 mentally retarded children (54.5%) at the school were in the 7–11 age group. Most respondents (85.5%) used an adaptive coping mechanism and a majority of them (60%) had a high level of stress.

The analysis revealed that, among the 47 respondents who used an adaptive coping mechanism, around 22 people (46.8%) experienced a low level of stress and around 25 people (53.2%) suffered from a high stress level. In the meantime, all eight respondents (100%) who used a maladaptive coping mechanism experienced a high stress level. The $p$-value of 0.006 indicated that there was a significant correlation between the coping mechanism being applied and the stress level of parents with mentally retarded children at SLB Wacana Asih, Padang (Table 1).

Table 1. Relationship between coping mechanism and stress level among parents with mentally retarded children in SLB Wacana Asih, Padang.

| | Stress level | | | | | | |
|---|---|---|---|---|---|---|---|
| | Low | | High | | Total | | |
| Coping mechanism | *f* | % | *f* | % | *f* | % | *p*-value |
| Adaptive | 22 | 46.8 | 25 | 53.2 | 47 | 100 | **0.006** |
| Maladaptive | 0 | 0 | 8 | 100 | 8 | 100 | |
| Total | 22 | 40 | 33 | 60 | 55 | 100 | |

## 4 DISCUSSION

A study by Jones and Passey (2005) on the adaptation to stress and the coping of 48 parents with disabled children in the UK showed that there was a significant relationship between coping and parenting stress. Mawardah et al. (2012) also reported that coping influenced the stress that was experienced by mothers of mentally retarded children. A similar study (Prasa, 2012) in Yogyakarta discovered that the stress experienced by parents originated both inside and outside the individuals, while the coping efforts detected included seeking social support, planned problem-solving,making plans self-control, distance, positive judgement, and acceptance of responsibility.

Of the 47 parents in our study who used adaptive coping mechanisms, 61.7% were in the age range 31–40 years old, and the children'saverage age ranged from 7 to 11 years old. A total of 39 (83%) parents had recent high school and college education. This level of education isreportedly quite high amongst Indonesian people. This better education level may lead to a better level of knowledge from which the parents can benefit. Suri and Daulay (2012) state that a person's educational level affects him/her because that level will determine how the individual faces his/her stressors. A higher education level will lead to better coping with stressors.

The results also showed that of the 47 parents using an adaptive coping mechanism, 43 (91.5%) were female and four (8.5%) were male, while all male parents using adaptive coping mechanisms had low stress levels. This finding suggests that male parents with mentally retarded children used the adaptive coping mechanisms more effectively, as indicated by how the coping mechanism decreased their stress level.

Our investigation of the 47 parents using an adaptive coping mechanism showed that 40 parents (85.1%) received emotional support from others, 37 parents (78.7%) sought peace in their religion, 32 respondents (68.1%) prayed about the circumstances that had happened to them, 33 parents (70.2%) tried to accept the condition of their children, and 38 parents (80.8%) tried to take care and find strategies to overcome the problems related to child mental retardation.

The results showed that as many as 25 (53.2%) parents experienced low stress levels, and 22 (46.8%) other parents experienced high levels of stress. From these results, it can be seen that the difference in the number of parents using adaptive coping mechanisms who experienced low stress levels and those with high stress levels is not very significant. This may be influenced by the fairly high level of education of most parents(88%) using adaptive coping mechanisms and having low stress levels. This finding supports the arguments of Suri and Daulay (2012).

More than half of the parents (76.7%) said that far more time and energy were consumedin taking care of a mentally retarded child, who is more dependent on the person closest to him/her, in this case the parent. In addition, a majority of parents (73.3%) also felt stressed and worried about the future of their disabled children due to the child's limitations and dependence on them. According to Johnson (in Sarafino & Smith, 2011), families need to adapt to the problem. To overcome various problems that can cause stress, individuals have different coping mechanisms. Triana and Andriany (2010) explained that parents respond to existing stressors in various ways, namely, acceptance, responsibility, seeking support, thinking of it as a life lesson, submitting to God, rejection, and sadness. Of these various responses, rejection and sadness are categorised as maladaptive coping mechanisms, while all other responses are considered to be adaptive coping mechanisms.

The data analysis showed that of the 55 parents, there were 30 adult carers with a high stress level and 25 other adult carers with a low stress level. Of the 30 parents who experienced high stress, the majority (96.7%) were female. According to Hayes and Watson. (2013), the impact of stress that occurs in families with children with backwardness suggests higher stress levels in the mothers. This high level of stress occurs because the mothers take the role of carer in the family (Zablotsky, et al., 2009). This study supported this argument by showing that four out of five (80%) male parents used an adaptive coping mechanism and had a low stress level. A stress parenting study conducted Jones, et al (2013) in parents of children with autism found that mothers experienced a higher level of stress than did fathers.

Of the 30 parents who experienced a high stress level, 22 used an adaptive coping mechanism and eight a maladaptive one. Of the 25 parents who experienced low stress levels, allused an adaptive coping mechanism. The effectiveness of a coping mechanism is confirmed when it can reduce the stress experienced by a person. If the coping being used is adaptive but does not reduce a person's stress level, then such coping is not effective (Hawari, 2011).

## 5 CONCLUSION

From this research, the researchers can conclude that most parents with mentally retarded children used adaptive coping mechanisms. More than half of the parents had high levels of stress. In 2016, there was a significant relationship between the coping mechanism being used and the parental stress level at SLB Wacana Asih, Padang.

It is suggested that the school provide direction to parents on utilising resources such as acceptance, positive outlook, and social support as adaptive coping mechanisms in dealing with child-related problems of mental retardation. Schools can also create sharing classes for parents with mentally retarded children in order to share information and support each other in handling the stress caused by mental retardation. For health services, it is hoped that access to mental health services for the parents is provided at a community level. The access may be in the form of family consultation services related to psychosocial problems experienced by families with mentally retarded children. Information about coping mechanisms to deal with stress in nurturing mentally retarded children should also be provided for the parents so that they can raise their disabled children and cope with the resulting stress in much better ways.

## REFERENCES


Bauman, S. (2004). Parents of children with mental retardation: coping mechanisms and support needs. Unpublished Dissertation. College Park, University of Maryland.

Carver, C.S. (1997). You want to measure coping but your protocol's too long: consider the brief COPE. *Int J Behav Med*. 4: 92–100.

Dardas, L.A & Ahmad, M.M. (2013). Coping strategies as mediator and moderators between stress and quality of life among parents of children with autistic disorder. *Stress Health*. 31, (5–12).

Davidson, G.C, Neale, J.M & Kring, A.M. (2006). Abnormal psychology. 9th ed. Jakarta, PT. Raja Grafindo Persada.

Hawari, D. (2011). Stress management, anxiety and depression. Jakarta, Balai Penerbit FKUI.

Hayes, S.A and Watson, S. L. (2013). The impact of parenting stress: a meta analysis of studies comparing the experience of parenting stress in parents of children with and without autism spectrum disorder. *Journal Autism Dev Disorder*. 43 (629–642).

Hidangmayum, N. & Khadi, P.B. (2012). Parenting stress of normal and mentally challenged children. *Karnataka Journal of Agricultural Sciences*, *25*(2), 256–259.

Islam, M.Z., Shahnaz, R. & Farjana, S. (2013). Stress among parents of children with mental retardation. *Bangladesh Journal of Medical Science*, *12*(1), 74–80.

Jones, J and Passey, J. (2005). Family adaptation, coping and resouces: parents of children with developmental disabilities and behavoir problems. Journal on Developmental Disabilities. 11, (31–46).

Jones, L, Totsika, V, Hastings, R.P, & Petalas, M. A. (2013). Gender differences whenparenting children with autism spectrum disorder: a multilevel modelling approach. *Journal of Autism and Developmental Disorder*. 43(9). 2090–2098.

Karasavvidis, S. (2011). Mental retardation and parenting stress. *International Journal of Caring Sciences*, *4*(1), 21–31.

Kosasih, E. (2012). *A wise way of understang children with special neees.* Bandung, Indonesia: YRama.

Lazarus, R.S and Folkman, S. (1984). Stress, appraisal and coping. New York, Springer.

Mangunsong, F. (2012). *Psychology and education for children with special needs: 2nd Ed.* Jakarta, Indonesia: LPSP3UI.

Manktelow, J. (2009). *Stress Control.* Jakarta, Indonesia: Erlangga.

Mawardah, U., Siswati & Hidayati, F. (2012). Relationship between active coping with parenting stress in mother of mentally retarded child. *Jurnal Psikologi Empati*, *1*(1), 1–14.

Ministry of Education and Culture. (2012). SLB Statistics. Jakarta, Indonesia: Ministry of Education and Culture of Republic of Indonesia.
Ministry of Health. (2014). Portrait of person with dissability. Jakarta, Ministry of Health of Republic of Indonesia
Nataya, R. (2014). Relationship between father social support and mother's anxiety level with mental retardation children *in SLBN 2Padang.* (Unpublished undergraduate thesis), Andalas University, Padang, Indonesia).
Prasa, B.A. (2012). Stress and coping in parents with mental retardation children (Unpublished undergraduate thesis, Ahmad Dahlan University, Yogyakarta, Indonesia).
Rasmun. (2004). Stress, coping and adaptation. Jakarta, CV. Sagung Seto
Sadock, B.J. & Sadock, V.A. (2007). *Kaplan & Sadock's synopsis of psychiatry: Behavioral sciences/ clinical psychiatry.* Philadelphia, PA: Wolters Kluwer.
Safaria, T. (2009). New understanding for meaningful life for parents of autistic children. Yogyakarta, Indonesia: Graha Ilmu.
Sarafino, E.P. & Smith, T.W. (2011). *Health psychology: Biopsychosocial interactions* (7th ed.). New York, NY: John Wiley & Sons.
Siddiqui, A.F. (2014). Sociodemographic profile of families with mentally retarded children and its relation to stress. *Bangladesh Journal of Medical Science*, *13*(4), 378–382.
Soemantri, S. (2007). *Psychology for special needs children*. Bandung, Indonesia: PT Refika Aditama.
Stuart, G.W. (2013). *Principles and practice of psychiatric nursing* (10th ed.). St Louis, Missouri: Elsevier Mosby.
Suri, D.P. & Daulay, W. (2012). Coping mechanisms for parents with Down syndrome children in SDLB Negeri 107708 Lubuk Pakam, Deli Serdang District. *Jurnal Universitas Sumatra Utara*, *1*(1), 52–56.
Swari, N. L. (2014). Relationship between coping mechanism and parenting style with mental retardation children at SLB. Journal of Nursing Science Udayana University. 2(2).
Triana, N.Y and Andriani, MM. (2010). Family stress and coping with mentally retarded child in SLB C and SLB C1 Widya Bhakti Semarang. Unpublished undergraduate Thesis. Diponegoro University, Semarang, Indonesia.
Videbeck, S. (2008). *Psychiatric nursing text book*. Jakarta, Indonesia: EGC.
Wenar, C. & Kerig, P. (2007). *Developmental psychopathology*. Singapore: McGraw-Hill.
Zablotsky, B, Bradshaw, C.P, & Stuart, E.A. (2013). The association between mental health, stress and coping support in mothers of children with autism spectrum disorder. *Journal of Autism and Developmental Disorder*. 43(6), 1380–1393.

*Medical and surgical*

*Strengthening Research Capacity and Disseminating New Findings in Nursing and Public Health – Malini et al. (Eds)*
 *ISBN 978-1-138-50066-2*

# Examining the effect of logotherapy on the anxiety levels of cervical cancer patients

A.R. Ma'rifah, M.M. Budi & R.I. Sundari
*Harapan Bangsa Institute of Health Sciences, Purwokerto, Indonesia*

ABSTRACT: Being a cervical cancer patient has many psychological impacts for the patient. Anxiety about the health condition can affect patient reluctance to seek examination and further treatment. This anxiety can be managed by health personnel, with one approach being logotherapy. This study examined the effect of logo-therapy on anxiety scores of cervical cancer patients using the Hamilton Anxiety Rating Scale (HARS). It was a quasi-experimental, pre- and post-test study, with a control group design, in which 42 patients were involved. The experimental group received logo-therapy. Analysis of covariance (ANCOVA) was used to analyse the post-test differences between the two groups, with covariates of the pretest. The results showed that the average anxiety of the experimental group reduced from 41.03 to 34.19, while that of the control group changed from 40.77 to 36.48. At alpha 0.05, the difference of anxiety in the treated group was significant (*p* value of 0.0407). The results of the statistical tests show that there is a significant difference in anxiety scores in the group receiving logo-therapy compared with the control group.

## 1 INTRODUCTION

Cervical cancer is noted as a deadly disease for women in Indonesia and is ranked second (10.3%) for hospitalised patients in Indonesia. The prevalence of cervical cancer in the United States is about 92 in 100,000 women with a fairly high mortality of 27 in 100,000, or 18% of deaths occurring in women. In Indonesia, 12 in 100,000 women, more than 80% of cases, are in an advanced stage of the cancer such that treatment to achieve healing is difficult (Depkes, 2012).

With so many cases, not everyone can deal with this condition. Reactions such as helplessness, despair, anxiety, depression or rebellion can dominate, so that the effects of additional symptoms and complications are more disturbing (Sjamsuhidayat R, Wim de Jong, 2004).

The disease process significantly impacts on the quality of life of patients and their families. Anxiety arises, then worsens in the critical phase during the illness and during treatment. In general, cancer patients feel anxiety when faced with a series of health and psychosocial problems involving the odds of life and death (Trill M.D, 2013).

Although cancer is a scourge, patients are expected to avoid excessive anxiety. The frequency and excessive anxiety can make cancer cells spread quickly (Andersen et al., 2008). One of the approaches that can be used to overcome anxiety is the use of psychopharmaceutics in the form of anti-anxiety drugs. The use of anti-anxiety medications can cause a complete depression of the central nervous system (Townsend, 2008).

Mavissakalian and Michelson (1986), in Videbeck (2008), suggests that psychotherapy is effective in overcoming anxiety disorders, especially when combined with pharmacotherapy. Another approach is the use of psychological therapy/psychotherapy. A spiritual system involving religious and cultural beliefs can help cancer patients retain life meaning when adjusting to a life-threatening illness (Evans, M., Shaw, A., & Sharp, D, 2012 ). Many studies have shown that the meaning of life and spirituality are essential parts in the experience of a person with serious illness.

Logo-therapy is an existential psychotherapy that focuses on the meaning of life. The impression of this meaning is obtained through the realisation of three types of 'value': (a) the creative value (what the individual gives to the world); (b) the value of experience (what the individual receives from the world); (c) attitudinal values (the ability to change one's attitude towards unchangeable circumstances) (Esping, 2011). Principally, logo-therapy can be integrated with techniques that are often used by health professionals (Schulenberg et al., 2008). The purpose of this study was to determine the effect of logotherapy on the anxiety levels of cervical cancer patients.

## 2 MATERIALS AND METHODS

The population in this study were mothers who had cervical cancer and attending Prof. Margono Soekarjo Hospital. A purposive sampling technique was used with 42 respondents being selected. The study was quasi-experimental using pre-tests and post-tests with a control group design.

Prior to the intervention, anxiety levels were measured. The Hamilton Anxiety Rating Scale (HARS) instrument was utilised for both groups. The intervention group was given logo-therapy comprising eight sessions according to Viktor Frankl's books and the articles of William Breitbart and colleagues, and adapted to the group's culture and religion. Meanwhile, the control group was given treatment in accordance with the nursing care standards of the hospital. Treatment was given for two weeks. Post-test measurements were made two days after the last intervention was administered.

## 3 RESULTS

Logotherapy group effectiveness in reducing the anxiety of women with cervical cancer was evaluated with analysis of covariance (ANCOVA). After calculating the pretest and post-test scores within the experimental and control groups, the results were then analysed by ANCOVA using SPSS 19 statistical software. The findings of this analysis are shown in Table 1.

According to Table 1, the pre-test score of the experimental group is higher than the post-test score on the anxiety scale. Levene's test was used in order to examine the equality hypothesis of variances. Furthermore, for determining the significance of these differences, ANCOVA was used.

As shown in Table 3, due to the significance level ($p < 0.05$), logotherapy had an effect in reducing anxiety ($F = 4.602$). The result of the present study determined that cervical cancer patients had experienced lower levels of anxiety through logotherapy intervention than those who had not participated in this intervention.

Table 1. Mean and standard deviation for cervical cancer patients in pretest and post-test scores ($n = 42$).

| | | Pretest | | Post-test | |
|---|---|---|---|---|---|
| Group | Var | Mean | SD | Mean | SD |
| Exp | Anxiety | 41.03 | 5.65 | 34.19 | 5.06 |
| Ctrl | Anxiety | 40.77 | 5.94 | 36.48 | 3.89 |

Table 2. Levene's test for homogeneity of variances.

| *F* | df1 | df2 | Sig. |
|---|---|---|---|
| 0.131 | 1 | 40 | 0.783 |

Table 3. Mean and standard deviation for cervical cancer patients in pretest and post-test scores ($n = 42$).

| Variance source | df | Sum of square | Mean square | $F$ | Sig. |
|---|---|---|---|---|---|
| Pretest | 1 | 39.404 | 39.404 | 1.893 | 0.000 |
| Group | 1 | 723.47 | 40.193 | 4.602 | 0.0407 |
| Error | 39 | – | 19.597 | – | – |

## 4 DISCUSSION

The results indicate that most of the respondents were in severe anxiety and panic. According to the medical data, most respondents were at stage IIb–IV cancer and had undergone repeated cycles of chemotherapy. Based on the interviews with the officer and the schedule of chemotherapy, the average patient will receive further chemotherapy for 2–3 months. In addition, some cases also present with other physical conditions such as anaemia, which requires them to stay in hospital longer due to blood transfusion requirements. This situation, coupled with the effects of chemotherapy perceived by patients as ranging from nausea and vomiting to hair loss, contribute to the psychological burden in the form of increased anxiety levels. Hidayat, AA (2011) proved that cervical cancer patients at an early stage IIa coming for treatment were 28.6%, the remains were 66.4% those at IIb–IVb stage were 37.3%, and those at stage IIIb were 37.3%. The data showed that patients came in already very late and looked for help only after bleeding, because in the early stages it often did not cause any symptoms. Whereas, cervical cancer at an advanced stage is incurable and has the course of chronic disease and deadly with regard to cervical cancer patients at an advanced stage should undergo a complex therapy for a long time.

The main psychological problem experienced by patients with advanced cervical cancer is psychological distress associated with the diagnosis of cancer, or physical and social problems. Patients' difficulties in accepting the state of pain will cause prolonged psychological distress, resulting in depression and poor cooperation, both in terms of treatment and maintaining a healthy body (Potter & Perry, 2005). Another study of psychological distress in 265 cancer patients at the start of chemotherapy showed that these conditions significantly affect anxiety and depression levels and degrade the quality of patients' lives (Iconomou et al., 2008).

In Table 3, it can be seen that logotherapy intervention can reduce the anxiety of cervical cancer patients. Logotherapy leads to a natural physiological response to stress. The function of logotherapy is to provide help to contrast the various human forms of 'metaphysical irresponsibility', gradually leading the patient towards a consciousness of the various possibilities and freedom implicit in his/her choices. Psychic distress, expressed through the neuroses, is seen as a 'falling' or 'stumbling' along a path that leads towards true meaning, and human suffering is considered not as a symptom, but rather as an 'action' or working condition (*Leistung*), which forms part of the dynamic of the spiritual decisions to be made during the course of one's life (Frankl,V.E. 2011).

According to Costantini et al. (2012), psychotherapy centred on the meaning of life and human dignity is the most significant approach in psycho-oncology. This study is also in accordance with the work of Bahar Garfami and colleagues who found that logo-therapy reduces depression and anxiety in breast cancer patients (Bahar Garfami et al., 2009). While Hamid et al. (2011) found that breast cancer patients participating in logotherapy experienced an increase in quality of life.

Speaking of the role of meaning in life, Jaarsma et al. (2007) noted that the experience of meaning in life is positively associated with feelings of psychological well-being and is negatively associated with feelings of distress. In addition, the work of Dezutter et al. (2013) showed that meaning in life may be related to the acceptance of one's condition.

## REFERENCES


Anagnostopoulos, F., Slater, J. & Fitzsimmons, D. (2010). Intrusive thoughts and psychological adjustment to breast cancer: Exploring the moderating and mediating role of global meaning and emotional expressivity. *Journal of Clinical Psychology in Medical Settings*, *17*(2), 137–149.

Andersen, B.L., Yang, H.-C., Farrar, W.B., Golden-Kreutz, D.M., Emery, C.F., Thornton, L.M., Carson, W.E. (2008). Psychologic intervention improves survival for breast cancer patients. *Cancer*, *113*, 3450–3458.

Bahar Garfami, H., Shafi Abadi, A. & Sanai Zaker, B. (2009). Effectiveness of group logo therapy in reducing symptoms of mental health problems in women with breast cancer. *Andisheh va Raftar*, *4*(13), 35–42.

Breitbart, W., Poppito, S., Rosenfeld, B., Vickers, A.J., Li, Y., Abbey, J., ... Cassileth, B.R. (2012). Pilot randomized controlled trial of individual meaning-centered psychotherapy for patients with advanced cancer. *Journal of Clinical Oncology*, *30*(12),1304–1309.

Costantini, A., Navarra, C., Brunetti, S. & Caruso, R. (2012). Psychotherapeutic interventions in psycho-oncology. *Neuropathological Diseases*, *1*(2), 145–160.

Dalimartha, S. (2008). *Care Your Self ( Ca Servik)*. Jakarta, Indonesia: Penebar Plus.

Depkes. (2012). *Indonesia Health Profile*. Jakarta, Indonesia: Ministry of Health, Republic of Indonesia. Indonesia Health Profile.

Dezutter, J., Casalin, S., Wachholtz, A., Luyckx, K., Hekking, J. & Vandewiele, W. (2013). Meaning in life: An important factor for the psychological well-being of chronically ill patients. *Rehabilitation Psychology*, *58*(4), 334–341.

Esping. (2011) Autoethnography as Logotheraphy; An Existensial Analysis of Meaningful Social Science Inquiry. *Journal Of Border Educational Research*.Vol.9 pp 59–67. Texas Christian University.

Evans, M., Shaw, A., & Sharp, D. (2012). Integrity in patients'stories:'Meaning making' through narrative in supportive cancer care. European Journal of cancer care. *European Journal of of Integrative Medicine,* 4 (1), e11–e18.

Frankl, V.E, (2011). *Logotheraphy Approach To Personality*. the United States International University. USSA.

Green, C.W. & Setyawati, H. (2005). *Small Book Series of Alternative Therapies*. Yogyakarta, Indonesia: Yayasan Spiritia.

Hamid, N., Talebiyan, L., Mehrabizade Honarmand, M. & Yavari, A.H. (2011). The effects of logotherapy on depression, anxiety and quality of life of cancer patients in Ahvaz Big Oil Hospital. *Journal of Psychological Achievements*, *4*(2), 199–224.

Hidayat Alimul, Aziz. (2011). *Methods of Nursing Research and Data Analysis Techniques.* Jakarta: Salemba Medika.

Iconomou Et Al., 2008. Emotional distress in cancer patients at the beginning of chemotherapy and its relation to quality of life.. https://www.ncbi.nlm.nih.gov/pubmed/18555468.

Jaarsma, T.A., Pool, G., Ranchor, A.V. & Sanderman, R. (2007). The concept and measurement of meaning in life in Dutch cancer patients. *Psycho-Oncology*, *16*(3), 241–248.

Kang, K.A., Im, J.I., Kim, H.S., Kim, S.J., Song, M.K. & Sim, S. (2009). The effect of logotherapy on the suffering, finding meaning, and spiritual well-being of adolescents with terminal cancer. *Journal of Korean Academy of Child Health Nursing*, *15*(2), 136–144.

Notoatmodjo, S. (2010).. Health Methodology Jakarta, Indonesia: Rineka Cipta.

Potter, P.A, Perry, A.G. Texbooks: Fundamentals of Nursing: Concepts, Processes, and Practices. Edisi 4. Volume 2. Alih Bahasa: Renata Komalasari, dkk. Jakarta: EGC. 2005.

Rosjidi, Imam. (20010). Epidemiology of cancer in Women Jakarta, Indonesia: Sagung Seto.

Schulenberg, S.E., Hutzell, R.R., Nassif, C. & Rogina, J.M. (2008). Logotherapy for clinical practice. *Psychotherapy: Theory, Research, Practice, Training*, *45*(4), 447–463.

Setiadi. (2013). Concepts and Practice of Nursing Research Writing (2nd ed). Yogyakarta, Indonesia: Graha Ilmu.

Sjamsuhidayat R, Wim de Jong, 2004. Textbook of Surgery, Edition 2, Jakarta: EGC.

Stuart, G.W. & Laraia, M.T. (2005). *Principles and practice of psychiatric nursing* (8th ed.). St Louis, MO: Mosby.

Stuart, G.W. (2013). *Principles and practice of psychiatric nursing* (10th ed.). St Louis, MO: Mosby Elsevier.

Townsend, M.C. (2008). *Essentials of psychiatric mental health nursing* (4th ed.). Philadelphia, PA: F. A. Davis Company.

Townsend, M.C. (2009). *Psychiatric mental health nursing* (6th ed.). Philadelphia, PA: F. A. Davis Company.

Trill M.D (2013). Anxiety and sleep disorders in cancer patients. https://www.ncbi.nlm.nih.gov/pmc/articles/PMC4041166/

Videbeck, S. L. (2008). Psychiatric mental helath nursing. (4th Ed). Philadelphia: Lippincott Williams & Wilkins.

*Strengthening Research Capacity and Disseminating New Findings in Nursing and Public Health – Malini et al. (Eds)*
*© 2018 Taylor & Francis Group, London, ISBN 978-1-138-50066-2*

# Relationship between duration of intravenous therapy and frequency of intravenous cannula replacement

Bayhakki
*Universitas Riau, Riau, Indonesia*

ABSTRACT: Most hospitalised patients receive intravenous therapy, with many factors affecting the frequency of intravenous cannula replacement. This study aimed to investigate the relationship between the duration of intravenous therapy and the frequency of intravenous cannula replacement. It was a correlational study with a cross-sectional approach. Forty-eight hospitalised patients who met the inclusion criteria were recruited. Data were collected using a questionnaire and analysed using the Spearman rank-order correlation test. The results showed a mean patient age of 44.21 years, and mean duration of intravenous therapy and frequency of intravenous cannula replacement of 3.9 days and 1.19 times, respectively. There was a significant relationship between duration of intravenous therapy and frequency of intravenous cannula replacement ($p$ = 0.000) with a Spearman correlation coefficient of 0.592, which indicated that the longer the duration of therapy, the more frequent the intravenous cannula replacement. Nurses are expected to carefully control the duration of intravenous catheter use to prevent problems such as phlebitis and pain due to intravenous cannula replacement.

*Keywords*: cannula, infusion, intravenous therapy, patients

## 1 INTRODUCTION

Patients undergoing hospitalisation usually receive intravenous (IV) therapies to facilitate healthcare workers in delivering medications, nutrition, blood and its products, and to monitor patient haemodynamics (Rickard et al., 2015). Peripheral and central venous cannulation are the most commonly used methods of intravenous therapy for hospitalised patients (Waitt et al., 2004). However, inserting peripheral cannulation carries a risk of causing complications in patients, such as phlebitis, pain, infiltration, extravasation and occlusion (Abolfotouh et al., 2014; Danski et al., 2016).

Phlebitis is the most common complication in patients receiving intravenous therapy during hospitalisation (Abolfotouh et al., 2014; Danski et al., 2016). Phlebitis in hospitalised patients can be caused by various factors, including duration of intravenous therapy, venous selection, use of an infusion pump, and the frequency of intravenous cannula replacement (Elvina & Kadrianti, 2014; Uslusoy & Mete, 2008). Previous studies have shown that patients receiving intravenous therapy over 72 hours are at risk of phlebitis (Danski et al., 2015). In general, hospitalised patients receive intravenous therapy for 48–120 hours; thus, they are at risk of developing phlebitis (Pasalioglu & Kaya, 2014). Previous research results in patients undergoing hospitalisation showed that as many as 43.3% of patients receiving intravenous therapy for ≥ 72 hours had phlebitis (Elvina & Kadrianti, 2014). Therefore, nurses have an important role in the management of intravenous therapy in patients to prevent complications (Uslusoy & Mete, 2008).

The Centers for Disease Control and Prevention (CDC) recommends replacing the catheter after 72–96 hours of intravenous therapy to prevent infection and phlebitis (O'Grady et al., 2011). However, previous studies have indicated that intravenous catheter replacement performed only at the time of onset of clinical symptoms does not increase the risk of phlebitis compared with routine intravenous catheter replacement performed after 72–96 hours of intravenous catheter insertion (Abolfotouh et al., 2014). Therefore, the researchers were interested to investigate the relationship correlation between the duration of IV therapy and the frequency of IV cannula replacement.

## 2 METHODS

A correlational study with a cross-sectional approach was performed. Forty-eight hospitalised patients who met the inclusion criteria were recruited. The inclusion criteria included being hospitalised for at least one day, able to communicate verbally, and willing to participate in the study. Information about age, sex, education, employment, length of time receiving intravenous therapy, and frequency of intravenous cannula replacement was collected using a questionnaire.

### 2.1 *Statistical methods*

Descriptive analysis was used to describe the demographic statistics of the respondents. Bivariate analysis was performed to assess the correlation between the duration of IV therapy and frequency of IV cannula replacement using the Spearman rank-order correlation test. A value of $p < 0.05$ was considered to be statistically significant.

## 3 RESULTS

Forty-eight hospitalised patients who met the inclusion criteria participated in this study. The mean of age of the patients was 44.21 years, and the mean duration of intravenous therapy and frequency of intravenous cannula replacement were 3.9 days and 1.19 times, respectively. Most of the patients were male (62.5%). The majority of the patients had education to senior high school level (43.8%). Most patients were housewives (33.3%) or entrepreneurs (33.3%) (see Table 1).

The results of the Spearman test show a significant relationship between length of time receiving intravenous therapy and frequency of intravenous cannula replacement ($p = 0.000$). The strength and direction of the correlation reflected a correlation coefficient of 0.592, which means the correlation is strong and positive, such that the longer the duration of IV therapy, the more frequent the replacement of the intravenous cannula (Table 2).

## 4 DISCUSSION

There was a significant relationship between the duration of intravenous therapy and the frequency of intravenous cannula replacement, which indicated that the longer the duration of therapy, the more frequent the intravenous cannula replacement. The duration of intravenous therapy in each hospitalised patient varied, but the longer the patient was given intravenous therapy the higher the patient's risk of complications (Purnamasari et al., 2013). In this study, the majority of patients had been receiving intravenous therapy for an average of four days with an intravenous catheter replacement frequency of one. The results are in line with the recommendations of the Centers for Disease Control and Prevention that intravenous catheter replacement should be performed after 72–96 hours of intravenous therapy to prevent infection and phlebitis (O'Grady et al., 2011). Previous studies have shown

Table 1. Demographic characteristics ($n$ = 48).

| Variable | F (%) | Mean ± SD |
|---|---|---|
| Age | | 44.21 ± 17.691 |
| Sex | | |
| Male | 30 (62.5) | |
| Female | 18 (37.5) | |
| Education | | |
| No education | 2 (4.2) | |
| Elementary school | 12 (25) | |
| Junior high school | 8 (16.7) | |
| Senior high school | 21 (43.8) | |
| University/Bachelor | 3 (6.3) | |
| Employment | | |
| Unemployed | 8 (16.7) | |
| Housewife | 16 (33.3) | |
| Student | 2 (4.2) | |
| Entrepreneur | 16 (33.3) | |
| Retired | 2 (4.2) | |
| Employee | 4 (8.3) | |
| Medical diagnosis | | |
| Internal disease | 34 (70.8) | |
| Surgery | 9 (18.8) | |
| Obstetric | 2 (4.2) | |
| Duration of intravenous therapy | | 3.9 ± 6.537 |
| Frequency of intravenous cannula replacement | | 1.19 ± 1.525 |

Table 2. Correlation between duration of intravenous therapy and frequency of intravenous cannula replacement ($n$ = 48).

| Variable | Correlation coefficient ($r$) | $p$ value |
|---|---|---|
| Duration of intravenous therapy | 0.592 | <0.001* |

*$p$ value < 0.05.

different results, indicating that intravenous catheters should be replaced when clinical indications are found to avoid replacing them too frequently. Based on the results of randomised controlled trials in 362 patients, there was no significant difference found between routine catheter replacements every three days and catheter replacement when clinical indications of phlebitis occurred (Rickard et al., 2010). One study found that the highest incidence rate of phlebitis occurred in catheters inserted for 48 hours or less, with a decreased rate in catheters inserted for 49–96 hours, and the lowest rate in catheters remaining in place for 97–120 hours (Pasalioglu & Kaya, 2014). In addition, catheter replacement due to clinical indications can save the use of the device, save the health worker time, and reduce the discomfort suffered by the patient due to the replacement of the intravenous catheter (Rickard et al., 2010). The difference between the results of this study, supported by CDC recommendations, and the results of some previous studies needs to be further investigated to make it clearer when nurses should replace IV catheters.

However, nurses have an important role in reducing the complications arising from intravenous therapy by recognising predisposing factors such as phlebitis, pain, extravasation, infiltration or occlusion, because complications arising from the insertions of intravenous therapy are among the indicators of care quality (Milutinovic et al., 2015). In addition, intravenous therapy management is part of independent nursing care, so nurses are expected to control the duration of intravenous therapy in patients to improve patient safety and nursing care quality.

## 5 CONCLUSIONS

There is a significant relationship between the duration of intravenous therapy and frequency of intravenous cannula replacement. Nurses are expected to carefully control the length of time patients use intravenous catheters by observing intravenous therapy daily to prevent problems such as phlebitis and pain due to repeated intravenous cannula replacement.

## REFERENCES

Abolfotouh, M.A., Salam, M., Bani-Mustafa, A., White, D. & Balkhy, H. (2014). Prospective study of incidence and predictors of peripheral intravenous catheter-induced complications. *Therapeutics and Clinical Risk Management*, *10*, 993–1001.

Danski, M.T.R., de Oliveira, G.L.R., Johann, D.A., Pedrolo, E. & Vayego, S.A. (2015). Incidence of local complications in peripheral venous catheters and associated risk factors. *Acta Paulista de Enfermagem, 28*(6), 517–523.

Danski, M.T.R., Johann, D.A., Vayego, S.A., de Oliveira, G.R.L. & Lind, J. (2016). Complications related to the use of peripheral venous catheters: A randomized clinical trial. *Acta Paulista de Enfermagem, 29*(1), 84–92.

Elvina & Kadrianti, E. (2014). Factors related to occurance of phlebitis in Labuang Baji General Hospital, *Journal of Health Sciences, 5*(1), 28–32.

Milutinovic, D., Simin, D. & Zec, D. (2015). Risk factor for phlebitis: A questionnaire study of nurses' perception. *Revista Latino-Americana de Enfermagem*, *23*(4), 677–684.

O'Grady, N.P., Alexander, M., Burns, L.A., Dellinger, E.P., Garland, J., Heard, S.O., … Saint, S. (2011). Summary of recommendations: Guidelines for the prevention of intravascular catheter-related infections. *Clinical Infectious Diseases*, *52*(9), 1087–1099.

Pasalioglu, K.B. & Kaya, H. (2014). Catheter indwell time and phlebitis development during peripheral intravenous catheter administration. *Pakistan Journal of Medical Sciences*, *30*(4), 725–730.

Purnamasari, I.P., Ismonah & Hendrajaya. (2013). Relationship between length of using infusion and phlebitis in Tugurejo General Hospital, Semarang. Research. Unpublished.

Rickard, C.M., McCann, D., Munnings, J. & McGrail, M.R. (2010). Routine resite of peripheral intravenous devices every 3 days did not reduce complications compared with clinically indicated resite: A randomised controlled trial. *BMC Medicine*, *8*(53), 1–10.

Rickard, C.M., Marsh, N.M., Webster, J., Gavin, N.C., McGrail, M.R., Larsen, E., … Playford, E.G. (2015). Intravascular device administration sets: Replacement after standard versus prolonged use in hospitalised patients—a study protocol for a randomized controlled trial (The RSVP Trial). *BMJ, 5*, 1–7.

Uslusoy, E. & Mete, S. (2008). Predisposing factors to phlebitis in patients with peripheral intravenous catheters: A descriptive study. *Journal of the American Academy of Nurse Practitioners, 20*, 172–180.

Waitt, C., Waitt, P. & Pirmohamed, M. (2004). Intravenous therapy. *Postgraduate Medical Journal, 80*, 1–6.

*Strengthening Research Capacity and Disseminating New Findings in Nursing and Public Health – Malini et al. (Eds)*
*© 2018 Taylor & Francis Group, London, ISBN 978-1-138-50066-2*

# Cardiovascular disease risk assessment based on laboratory versus non-laboratory tests: A systematic review

D. Prabawati
*Sint Carolus School of Health Sciences, Jakarta, Indonesia*

J. Lorica
*St. Paul University Philippines, Tuguegarao, Philippines*

ABSTRACT: Cardiovascular Disease (CVD) risk assessment was applied to identify individuals who are more likely to develop CVD. The purpose of this review was to summarise the current literature on CVD risk assessment tools developed for adult patients using laboratory- and non-laboratory-based tests. A systematic review of the literature was conducted. The databases searched were ProQuest, EBSCO, ScienceDirect and PubMed, with a search period of 2011–2016. Fifteen randomised controlled trial studies met the criteria of this review. There were many laboratory-based CVD risk tools, with the Framingham Risk Score often considered the reference standard. Some studies discovered that non-laboratory-based CVD risk scores, when compared to laboratory-based scores, similarly ranked individuals and characterised CVD risk. The outcome of this review provides significant information that may be considered for CVD risk assessment that is cheaper and easier to apply in a large research population.

*Keywords*: cardiovascular risk; assessment; laboratory-based test, non-laboratory based test

## 1 INTRODUCTION

In the US, the Joint National Committee VII (JNC 7) report developed a new classification of blood pressure for adults aged 18 years or older, called prehypertension, because these individuals are at increased risk of progression to hypertension and show an independent increase in risk of Cardiovascular Disease (CVD) (Chobanian et al., 2003). Prehypertension is not a disease entity, rather it is a reminder to intervene in the patients' lifestyle immediately in order to prevent its progression to hypertension.

Early detection and treatment of individuals at CVD risk is an important strategy for preventing or delaying CVD events. It is expected that primary prevention, in terms of risk assessment, is needed in order to accurately determine and intervene early in the disease. CVD risk assessment is applied to identify individuals who are more likely to develop CVD and therefore receive more intensive interventions as soon as possible. Several tools have been developed for risk estimation, most of which are derived from the Framingham study. The majority of the risk assessment tools were applied according to laboratory values such as total cholesterol, High-Density Lipoproteins (HDLs), Low-Density Lipoproteins (LDLs), or other biomarkers. However, some tools have been developed which are non-laboratory-based, where the same modelling principles and assessment techniques were applied (D'Agostino et al., 2008).

Indonesia, like many other countries in South-East Asia, is experiencing a burden from CVD. Approximately one-third of the adult population have high blood pressure and nearly

1.5 million deaths occur due to hypertension each year in the South-East Asia region (Mohan et al., 2013). There are unhealthy behavior such as tobacco smoking (67% for males) and alcohol consumption that may lead to high blood pressure and obesity that has close relation for the development of CVD event or any non-communicable diseases Indonesia (World Health Organization, 2014).

Since 2014, the government of Indonesia has introduced various social insurance programmes for health; however, the focus is on curative care services and health infrastructure that supports medical care (Asia Pacific Observatory on Health Systems and Policies, 2017). Thus, the allocation for prevention is relatively low. Therefore, the resources for conducting research on CVD assessment with laboratory-based tests will be insufficient and, thus, some researchers specifically use non-laboratory-based tests as their tools (Ng et al., 2006; Dewi, Stenlund, Hakimi, & Weinehall, 2015).

## 2 METHOD

A systematic review was conducted to learn more and develop an analysis of literature published about CVD risk assessments used in various settings. The author searched the literature individually and identified many relevant studies; as a result, the quality of the studies reviewed and included can vary widely. This systematic review followed the guidelines of the Preferred Reporting Items for Systematic Reviews and Meta-Analyses (PRISMA).

As relevant publishers of journals related to CVD risk, the electronic databases searched were ProQuest, EBSCO, ScienceDirect and PubMed. The search period was 2011–2016. Keywords for the search were 'cardiovascular disease', 'risk', 'tool' and 'assessment'. The inclusion criteria in this study were: the article must be written in English, peer-reviewed, have free full text and published in a scholarly journal. The initial search generated a total of 1,050 papers from all search databases. After further filtering, as shown in Figure 1, a total of 15 articles were identified for in-depth screening and analysis in this study.

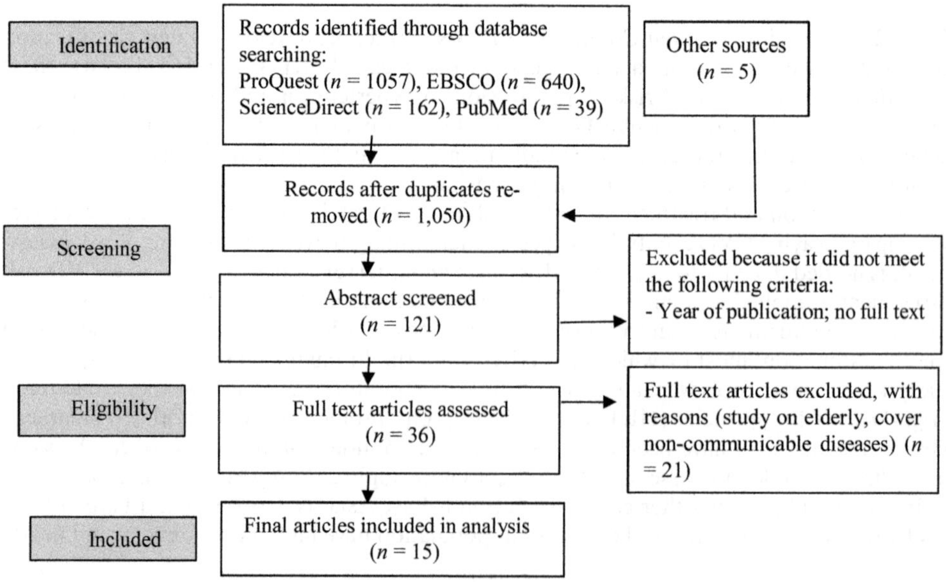


Figure 1. Diagram of information flow through phases of the systematic review (The PRISMA Group, 2009).

## 3 RESULTS

The following are the results of the literature search for relevant articles on primary CVD risk assessments/instruments. Most of the studies were excluded at the abstract stage because either the study was not relevant to the topic of CVD (coronary heart disease or myocardial infarction events) or the study population involved older adults and covered non-communicable diseases. There were some studies conducted in Asia and also in Indonesia; however, CVD risk assessment for Asian people is unmet. Most of the instruments originated in Europe or America.

Some studies in South-East Asia applied CVD risk estimation in their research, with most of them using the Framingham Risk Score (FRS) as their tool (Su et al., 2015; Singapore Ministry of Health, 2011). Moreover, Bitton and Gaziano (2010) stressed that only FRS has been evaluated in a developing country (China). The traditional independent risk factors for CVD used were abnormal lipid levels and blood sugar as a laboratory complement. However, using laboratory-based CVD risk assessment for screening purposes will be far too costly in most developing country settings; limited resources mean that they are unlikely to be adopted as policy in these settings. This has prompted an investigation into the possibility of using other known CVD risk factors that are easier and less costly to measure in place of the standard CVD risk factors that require costly laboratory tests when calculating CVD risk scores. It was confirmed that non-laboratory-based screening generated results similar to laboratory-based methods.

The results showed the summary of CVD risk assessment tools for non-laboratory-based. Some CVD risk assessment tools still need laboratory input, such as the World Health Organization's STEPwise; however, the laboratory needed is only simple and the blood sample can be drawn peripherally from a finger. In the study conducted by Dewi, Stenlund, Hakimi, & Weinehall (2015), which used STEPwise as a tool, they failed to conduct STEP 3 for biochemical measurement due to financial constraints. Nevertheless, the findings showed a positive change in CVD risk factors over a five-year period.

Body Mass Index (BMI), as measured in CVD risk tools, can be grouped into categories of underweight, normal, overweight and obese. The categories of obesity and overweight have a close relation with elevated triglyceride levels. It is known that high triglyceride levels may contribute to hardening of the arteries or thickening of the artery walls and, thus, may result in an individual being at high risk of Coronary Heart Disease (CHD) events. Typically, the obese and overweight also have low HDL cholesterol levels, with HDLs appearing to protect blood vessels from atherogenesis. In addition, the National Institutes of Health (2002) argue that HDL levels at levels above 35 mg/dl predict CHD, and the low HDL cholesterol associated with obesity and physical inactivity causes development of CHD states.

## 4 DISCUSSION

The Framingham risk assessment was updated by D'Agostino et al. (2008) to predict all CVD events, such as CHD, stroke, peripheral vascular disease and heart failure. In addition, they found that BMI could be substituted for cholesterol and HDLs when laboratory data were not available due to cost limitations without loss of predictive discrimination of the risk score. Some researchers contend that an advantage of the BMI model is that it can be used in low-resource settings without access to a laboratory. BMI-based CVD risk scores might have additional benefits in insured health-plan populations, as they could be used to identify people at high or moderate risk of CVD without laboratory values, and who might benefit from further assessments, as well as those at low risk whose risk assessment would not change based on additional laboratory testing. Green et al. (2015) highlighted that, overall, the correlation between non-laboratory- and laboratory-based risks was good, particularly among those at low risk for CVD, people who were younger, female, without diabetes and not on medications to lower blood pressure or lipids.

Faeh et al. (2012) discussed BMI versus cholesterol in CVD risk prediction models, revealing that a BMI model showed higher risk at all ages and could better discriminate individuals at high and low CVD risk. Moreover, they highlighted that using BMI instead of cholesterol in CVD risk prediction models may provide more accurate estimates (Green et al., 2015). Furthermore, lifestyle changes for diet and physical activity, promoting weight loss or preventing weight gain, may improve health more effectively than lipid-lowering treatment.

Gaziano et al. (2013) discovered that a non-laboratory-based CVD risk score, when compared to laboratory-based Framingham Risk Score (FRS), The European SCORE risk score and Italian CUORE project risk score ranked individuals similarly and characterised CVD risk. They also claimed that there was a strong agreement in risk characterisation between the non-laboratory- and laboratory-based scores in all cohorts; however, the correlation at CVD risk thresholds of >20% was highest for the laboratory-based CVD outcome score.

The General Practice model was applied to women aged 60–79 years, who were free form CHD and CVD events. The self-rated general health was obtained through interview or self-completed questionnaire. However, this model may not be appropriate in predicting CVD risk for a young adult population (May, Lawlor, Brindle, Patel & Ebrahim, 2006).

## 5 CONCLUSION

Despite the limitations, this systematic review discovered a number of CVD risk assessment tools for laboratory- and non-laboratory-based use for CVD risk and hypertension patients in studies conducted in developed countries. This review also revealed the weaknesses and strengths of laboratory- and non-laboratory-based tests for CVD risk assessment, although some studies confirmed that non-laboratory-based screening generated results similar to laboratory-based methods. Thus, non-laboratory-based testing could be an alternative for developing countries. Based on the findings, future studies on CVD risk assessment that are suitable and appropriate for Asian society need to be conducted and reviewed. However, the challenges in this review suggest that it is important to conduct studies of developing countries, so that those who seek to plan and implement CVD risk assessment could be much better prepared.

## REFERENCES

Asia Pacific Observatory on Health Systems and Policies. (2017). The Republic of Indonesia health system review. *Health Systems in Transition, 7*(1), 1–292.

Bitton, A., & Gaziano, T. (2010). The Framingham heart study's impact on global risk assessment. *Progress in Cardiovascular Diseases, 53*(1), 68–78.

Chobanian, A.V., Bakris, G.L., Black, H.R., Cushman, W.C., Green, L.A., Izzo, J.L., Jr., ... Roccella, E.J. (2003). The seventh report of the Joint National Committee on Prevention, Detection, Evaluation, and Treatment of High Blood Pressure: The JNC 7 report. *Journal of the American Medical Association, 289*(19), 2560–2571. doi:10.1161/01.HYP.0000107251.49515.c2.

D'Agostino, R.B., Vasan, R.S., Pencina, M.J., Wolf, P.A., Cobain, M., Massaro, J.M. & Kannel, W.B. (2008). General cardiovascular risk profile for use in primary care: The Framingham heart study. *Circulation, 117*(6), 743–753. doi:10.1161/CIRCULATIONAHA.107.699579.

Dewi, F. S. T., Stenlund, H., Hakimi, M., & Weinehall, L. (2015). An increase in risk factors for cardiovascular disease in Yogyakarta, Indonesia: a comparison of two cross-sectional surveys. *Southeast Asian Journal of Tropical Medicine and Public Health, 46*(4), 775–785.

Faeh, D., Braun, J., & Bopp, M. (2012). Body mass index vs cholesterol in cardiovascular disease risk prediction models, *172*(22), 2012–2014.

Gaziano, T.A., Pandya, A., Steyn, K., Levitt, N., Mollentze, W., Joubert, G., ... Laubscher, R. (2013). Comparative assessment of absolute cardiovascular disease risk characterization from non-laboratory-based risk assessment in South African populations. *BMC Medicine, 11*(1), 170. doi:10.1186/1741-7015-11-170.

Gaziano, T.A., Young, C.R., Fitzmaurice, G., Atwood, S. & Gaziano, J. M. (2008). Laboratory-based versus non-laboratory-based method for assessment of cardiovascular disease risk: The NHANES I follow-up study cohort. *The Lancet, 371*(9616), 923–931. doi:10.1016/S0140-6736(08)60418-3.

Green, B. B., Anderson, M. L., Cook, A. J., Catz, S., Fishman, P. A., McClure, J. B., ... Karmali, K. (2015). Using body mass index data in the electronic health record to calculate cardiovascular risk. *American Journal of Preventive Medicine*, *42*(4), 342–347. https://doi.org/10.1371/journal.pone.0119183.
Mohan, W., Seedat. Y.K. & Pradeepa, R. (2013). The rising burden of diabetes and hypertension in Southeast Asian and African regions: Need for effective strategies for prevention and control in primary health care settings. *International Journal of Hypertension, 2013*, 409083. doi:10.1155/2013/409083.
Moher D, Liberati A, Tetzlaff J, Altman DG, The PRISMA Group (2009). Preferred Reporting Items for Systematic Reviews and Meta-Analyses: The PRISMA Statement. PLoS Med 6(6): e1000097. doi:10.1371/journal.pmed1000097.
National Institutes of Health. (2002). Third report of the National Cholesterol Education Program (NCEP) expert panel on detection, evaluation, and treatment of high blood cholesterol in adults (Adult Treatment Panel III) final report. *Circulation*, *106*(25), 3143–3421.
Ng, N., Stenlund, H., Bonita, R., Hakimi, M., Wall, S. & Weinehall, L. (2006). Preventable risk factors for noncommunicable diseases in rural Indonesia: Prevalence study using WHO STEPS approach. *Bulletin of the World Health Organization*, *84*(4), 305–313.
Singapore Ministry of Health. (2011). *Screening for cardiovascular disease and risk factors.*
Su, T.T., Amiri, M., Mohd Hairi, F., Thangiah, N., Bulgiba, A. & Majid, H.A. (2015). Prediction of cardiovascular disease risk among low-income urban dwellers in metropolitan Kuala Lumpur, Malaysia. *BioMed Research International, 2015*, 516984. doi:10.1155/2015/516984.
World Health Organization. (2003). *The WHO STEPwise approach to noncommunicable disease risk factor surveillance (STEPS)*. Geneva, Switzerland: World Health Organization.
World Health Organization. (2014). *Non communicable diseases country profiles* (pp. 1–210). Geneva, Switzerland: WHO Press.

*Strengthening Research Capacity and Disseminating New Findings in Nursing and Public Health – Malini et al. (Eds)*
*© 2018 Taylor & Francis Group, London, ISBN 978-1-138-50066-2*

# Quality of life of cervical cancer patients post chemotherapy

E.A.F. Damayanti
*Lambung Mangkurat University, Banjarbaru City, South Kalimantan, Indonesia*

H. Pradjatmo
*Gadjah Mada University and RSUP Dr. Sardjito, Yogyakarta, Indonesia*

W. Lismidiati
*GadjahMada University, Yogyakarta, Indonesia*

ABSTRACT: Increasing the number of cervical cancer's survivor, it gives a better motivation in research field to escalate the impact of the disease and its treatment on patient's quality of life. This study was conducted to determine the quality of life in patients with cervical cancer, post chemotherapy. The study population included 60 post-chemotherapy cervical cancer patient sat the Dr. Sardjito (*RSUP Dr. Sardjito*) and Dr. Kariadi (*RSUP Dr. Kariadi*) Hospitals in 2016, selected by consecutive sampling. This study used the Indonesian version of the European Organization for Research and Treatment of Cancer Quality of Life Questionnaire, Core 30 (EORTC QLQ-C30). Patients' quality of life with cervical cancer post chemotherapy was found to have a functional level and health status that was quite high with a mean value on the functional scale above 50, and amean health status score of 59.98 ± 15.116; in addition, the level of symptoms was fairly low with a mean symptom scale value below 50, with the exceptions off atigue, nausea and vomiting, and loss of appetite. Therefore, management of treatment side effects is the most important aspectin improving overall quality of life for such patients.

## 1 INTRODUCTION

Cervical cancer is one of the most common cancers affecting women. The World Health Organization (WHO) noted that 85% of women develop cervical cancer in developing countries, including Indonesia. WHO data show that about 15,000 cases of cervical cancer are found in Indonesia every year. Indonesia is the country with the second highest number of cervical cancer cases in the world: every day, on average, 40 women are diagnosed with cervical cancer and 20 of them diefrom cervical cancer (Globocan, 2008). It is estimated that approximately one third of cases of cancer, including cervical cancer, come to the health services in advanced stages where the cancer has spread to other organs (Indonesian Cancer Foundation, 2013).

Increasing the number of cervical cancer's survivor, it gives a better motivation in research field to escalate the impact of the disease and its treatment on patient's quality of (Zeng et al., 2011). Cancer affects many dimensions of health and well-being. Ideally, treatment should not only extend survival and the disease-free period, but also reduce symptoms of the disease rather than cause side effects, and enhance the ability of individuals to return to a normal life (King & Hinds, 2012). Therapeutic treatment options for cervical cancer include surgical therapy, chemotherapy and radiotherapy. The side effects that arise due to chemotherapy may affect the physical, psychological, social and spiritual aspects of the patient so as to lower the quality of life of those with cervical cancer. Based on these side effects, we conducted this study to determine the quality of life in patients with cervical cancer, post chemotherapy.

## 2 METHOD

This study used a cross-sectional design. It was conducted at Dr. Sardjito Hospital and Dr. Kariadi Hospital between 15 July 2015 and 15 January 2016. The study population were all post-chemotherapy cervical cancer patients at these hospitals during this period. The study used a total sampling technique with 60 post-chemotherapy cervical cancer patients as respondents. The inclusion criteria were: patients with advanced cervical cancer, in the range IIB–IVB; patients aged between 40 and 60 years; patients having anBMI (Body Mass Index) 18.5; patients having completed one cycle of chemotherapy; and patients willing to become respondents in this study. Exclusion criteria were post-chemotherapy cervical cancer patients that could not attendone week after chemotherapy or lived more than 40 km away from Dr. Sardjito and Dr. Kariadi Hospitals.

The research instrument used in this study was the European Organization for Research and Treatment of Cancer Quality of Life Questionnaire, Core 30 (EORTC QLQ-C30), version 3.0, in the Indonesian language. Measuring the quality of life of patients was done within one week of completing the first cycle of chemotherapy. This study was approved by the ethical committee of the Biomedical Research Medicine Faculty, Gadjah Mada University.

## 3 RESULTS

### 3.1 *Characteristics of respondents*

The characteristics of the respondentsare shown in Table 1.

Table 1 shows that there were 39 respondents (65%) aged 51–60 years; 40 respondents (66.7%) had a good nutritional status; 47 respondents (78.3%) were married women; and the highest education of 35 respondents (58.3%) was elementary school; 47 respondents (78.3%)

Table 1. Characteristics of post-chemotherapy respondents ($n = 60$).

| Characteristic | $n$ (%) |
|---|---|
| Age | |
| 40–50 years | 21 (35) |
| 51–60 years | 39 (65) |
| Nutritional status | |
| Good | 40 (66.7) |
| Excess | 20 (33.3) |
| Marital status | |
| Married | 47 (78.3) |
| Widow | 13 (21.7) |
| Education | |
| Elementary school | 35 (58.3) |
| Middle school | 13 (21.7) |
| High school | 10 (16.7) |
| Diploma | 0 (0) |
| Bachelor | 2 (3.3) |
| Cervical cancer stage | |
| IIB | 14 (23.3) |
| IIIA | 0 (0) |
| IIIB | 44 (73.3) |
| IVA | 1 (1.7) |
| IVB | 1 (1.7) |
| Medications | |
| Ondansetron | 60 (100) |
| No medications | 0 (0) |

Table 2. Quality of lifeof post-chemotherapy cervical cancer patients ($n = 60$).

| Quality of life | Mean ± SD |
|---|---|
| Functional scale | |
| a. Physical function | 71.05 ± 23.765 |
| b. Role function | 64.55 ± 31.624 |
| c. Emotional function | 82.38 ± 18.831 |
| d. Cognitive function | 86.32 ± 16.657 |
| e. Social function | 78.07 ± 25.160 |
| General health scale | 59.98 ± 15.116 |
| Symptom scale | |
| a. Weakness | 51.57 ± 22.598 |
| b. Nausea and vomiting | 65.58 ± 24.785 |
| c. Pain | 34.97 ± 30.744 |
| d. Breathing difficulties | 9.95 ± 18.683 |
| e. Insomnia | 39.43 ± 35.066 |
| f. Loss of appetite | 63.90 ± 34.952 |
| g. Bowel problems | 37.25 ± 42.602 |
| h. Diarrhoea | 8.88 ± 23.682 |
| i. Financial difficulties | 37.15 ± 33.731 |

were housewives; 44 respondents (73.3%) had stage IIIBcervical cancer. All 60 of the post-chemotherapy respondents/patients (100%) received ondansetron medication.

### 3.2 *Quality of life of cervical cancer patients, post chemotherapy*

The results for quality of life of post-chemotherapy cervical cancer patientsare shown in Table 2.

The mean values on the functional scale for post-chemotherapy patients are highest in cognitive function and lowest in role function. The mean value on the general health scale for post-chemotherapy patients is quite high, with a value of 59.98 ± 15.116. The mean values on the symptom scale for post-chemotherapy patients are highest for nausea and vomiting, and loss of appetite, while the lowest mean values involved breathing difficulties and diarrhoea.

## 4 DISCUSSION

Chemotherapy is one therapy modality commonly used to treat cervical cancer. This treatment option has significant side effects on most bodily systems, which causes an impact on the patient's quality of life. Awareness of the importance of quality of life measurements for cervical cancer patients is key to managing the side effects (Davidson, 2011).

The results of our study indicate that the functional levels and general health of respondent post-chemotherapy cervical cancer patients were quite good. This is consistent with the study of Sulistyowati et al. (2006) regarding patients' quality of life with chemotherapy, which showed that most respondents have a good quality of life. Our study also shows that on the symptom scale post-chemotherapy patients scored quite low, with the exception of two symptoms with higher mean values.

One of the manifestations of chemotherapeutic agents is digestive tract problems. The most common side effects of chemotherapy are nausea and vomiting (Di Saia & Creasman, 2012). Symptoms of nausea and vomiting can cause physiological and psychological discomfort, affecting the quality of life of patients receiving chemotherapy. Factors associated with chemotherapy that may affect the occurrence and severity of symptoms of nausea and vomiting include the specific chemotherapy agents used, and their dosage, schedule and route of administration (Skeel & Khleif, 2011). This is consistent with the research of Must of a (2012), which compared the quality of life of patients with cervical cancer who received cisplatin-based chemotherapy The results showed that the problems most frequently complained of by patients included fatigue, nausea and vomiting, pain, loss of appetite, and financial difficulties.

Perwitasari and colleagues showed that 74.9% of 179 patients experienced symptoms of vomiting over the five days after chemotherapy despite using anti-emetic drugs. Patients with gynaecological cancers in the Perwitasari study experienced symptoms after chemotherapy of fatigue, nausea, vomiting, loss of appetite, and pain. The symptoms of nausea and vomiting experienced by these patients had a negative effect on their quality of life (Perwitasari et al., 2011).

Most chemotherapeutic agents indicated for cervical cancer have side effects of nausea and vomiting ranging from mild to severe. A chemotherapeutic agent that can cause severe emesis is cisplatin and cyclophosphamide in a high dose. Our study shows that the level of symptoms of cervical cancer patients post chemotherapy is quite high in relation to nausea and vomiting, and loss of appetite, which both have a mean value above 50.Although all cervical cancer patients undergoing chemotherapy receivedanti-nausea medication such as ondansetron, it was found that the side effects of chemotherapy on the digestive tract were severe enough to cause an impact on patients' quality of life. One strategy that can be used to manage the side effects of chemotherapy on the digestive tract is nutritional management. Some studies have provided strong evidence that nutritional intervention improves nutritional outcomes such as dietary intake and weight (Isenring et al., 2013).

Therefore, management of treatment side effects is the most important aspect of improving tolerance of therapy and overall quality of life (Skeel & Khleif, 2011).One management strategy in the treatment of cervical cancer patients with chemotherapy is the implementation of palliative care to relieve side effects and improve quality of life. Chemotherapy tolerability needs to be considered to maintain quality of life, especially for advanced cervical cancer patients (Davidson, 2011).

## 5 CONCLUSION

The quality of life of post-chemotherapy cervical cancer patients in terms of level and general health was pretty good; in addition, symptoms were quite low except for fatigue, nausea and vomiting, and loss of appetite. Good quality of life instrument shall be provided to measure the successful of nursing care patients with cervical cancer post chemotherapy.

## ACKNOWLEDGEMENT

The authors acknowledge the research assistant for data retrieval and participants of this study.

## REFERENCES

Davidson, S. (2011). Treatment for advanced cervical cancer: Impact on quality of life. *Critical Review in Oncology/Hematology*, 79, 24–30.

Di Saia, P.J. & Creasman, W.T. (Eds.). (2012). *Clinical gynecologic oncology* (8th ed.). Philadelphia, PA: Elsevier Saunders.

Indonesian Cancer Foundation. (2013, October). *Press Release TOT: Training of Trainers Pap Tes dan IVA Yayasan Kanker Indonesia (Indonesian Cancer Fondation)*. Jakarta, Indonesia: YayasanKanker Indonesia. Retrieved from https://www.facebook.com/kankerindonesia/posts/506094629486926.

International Agency for Research on Cancer. Globocan 2008 fast stats [Internet] Lyon, France: International Agency for Research on Cancer; 2010. [cited 2012 Jun 11]. Available. from: http://globocan.iarc.fr/factsheets/populations/factsheet.asp?uno=900.

Isenring, E., Zabel, R., Bannister, M., Brown, T., Findlay, M., Kiss, N., … Bauer, J. (2013). Updated evidence-based practice guidelines for the nutritional management of patients receiving radiation therapy and/or chemotherapy. *Nutrition & Dietetics*, *70*, 312–324.
King, C.R. & Hinds, P.S. (2012). *Quality of life: From nursing and patient perspectives*. Burlington, MA: Jones and Bartlett Learning.
Mustofa, I.M. (2012). *Perbandingan Kualitas Hidup Penderita Kanker Serviks yang Diberikan Kemoradiasi Berbasis Cisplatin dengan Cisplatin Tunggal di RSUP Dr. Sardjito Yogyakarta* (Thesis, Obstetrics & Gynaecology Department, Universitas Gadjah Mada Medicine Faculty and Dr. Sardjito Hospital, Yogyakarta, Indonesia).(Comparative study of quality of life servical cancer patient with Cisplatin-radiation combined and Cisplatin Single used only in *RSUP Dr. Sardjito Yogyakarta)*.
Perwitasari, D.A., Atthobari, J., Mustofa, M., Dwiprahasto, I., Hakimi, M., Gelderblom, H., ... Kaptein, A.A. (2011). Impact of chemotherapy-induced nausea and vomiting on quality of life in Indonesian patients with gynecologic cancer. *International Journal of Gynecological Cancer,* 22(1), 139–45.
Skeel, R.T. & Khleif, S.N. (2011). *Handbook of cancer chemotherapy* (8thed.). Philadelphia, PA: Lippincott Williams and Wilkins.
Sulistyowati, Y.N., Widyawati, W. & Aulawi, K. (2006). Kualitas Hidup Penderita Karsinoma Serviksdengan Kemoterapi di RS Dr. Sardjito Yogyakarta. *Jurnal Ilmu Keperawatan, 1*(3), 131–138. (The quality of life patient with cervical carcinoma retrieved chemotherapy in Dr. Sardjito Yogyakarta).
Zeng, Y.C., Ching, S.S.Y. & Loke, A.Y. (2011). Quality of life in cervical cancer survivors: A review of the literature and directions for future research. *Oncology Nursing Forum*, *38*(2), E107–E116.

*Strengthening Research Capacity and Disseminating New Findings in Nursing and Public Health – Malini et al. (Eds)*
*© 2018 Taylor & Francis Group, London, ISBN 978-1-138-50066-2*

# Community knowledge of coronary heart disease in relation to characteristic risk factors

E.A. Sari & S.H. Pratiwi
*Faculty of Nursing, Universitas Padjadjaran, Bandung, West Java, Indonesia*

ABSTRACT: This study aimed to determine the level of community knowledge of coronary heart disease in relation to characteristic risk factors. It used a descriptive, quantitative approach. Participants consisted of 70 people in Limusgede village in Cimerak district, Pangandaran regency, West Java, Indonesia, acquired through purposive sampling. Knowledge was measured using a questionnaire, while risk factor data were determined by age, body mass index, blood pressure, smoking behaviour, diabetes mellitus and physical activity level, which were referenced by Jakarta Cardiovascular Scores. The results showed that most of the respondents (61.43%) did not have good knowledge of coronary heart disease and the majority of respondents (87.15%) had moderate- or high-risk factors for cardiovascular disease. Further, almost half the respondents who had inadequate knowledge also had moderate- and high-risk factors for cardiovascular disease. Inadequate knowledge in people with moderate- or high-risk factors for cardiovascular disease should be a concern for health professionals.

## 1 INTRODUCTION

One of the main contributors to the mortality rate in the world is coronary heart disease. The disease is caused by a blockage or constriction of the coronary arteries due to the process of atherosclerosis or spasm or a combination of both (Smeltzer et al., 2010).

Coronary heart disease is affected by several risk factors including modifiable and non-modifiable risk factors. The risk factors that can be modified are hyperlipidemia, hypertension, smoking, diabetes mellitus, obesity, physical inactivity, alcohol consumption, stress, and high fat, cholesterol and calorie diets, while the risk factors that cannot be modified are age, gender and family history. The presence of these risk factors in a person can mean that the person is more likely to develop coronary heart disease compared to those who do not have these risk factors (Indonesian Ministry of Health, 2011).

To reduce the prevalence of coronary heart disease, preventive measures such as primary prevention and secondary prevention are required. This can be done by increasing the awareness of people in identifying risk factors and undertaking efforts to prevent and mitigate such risk factors, especially modifiable risk factors (Indrawati, 2014)

People who have high or moderate risk factors for coronary heart disease should practice good preventive behaviour, but this also requires a good level of knowledge so that the expected behaviour can be performed. In other words, a good level of knowledge and favourable behaviour are important factors in the efforts to prevent coronary heart disease (Dalusung-Angosta, 2010). There is a correlation between knowledge and the ability of coronary heart disease patients to avoid or mitigate risk factors (Indrawati, 2014).

This study aimed to determine the level of community knowledge of coronary heart disease in terms of the characteristic risk factors.

## 2 METHODS

This research used a descriptive, quantitative approach. Participants consisted of 70 people in Limusgede village in Cimerak district, Pangandaran regency, West Java, Indonesia, acquired through a non-probability technique of purposive sampling. Knowledge was measured using a questionnaire, while risk factor data were determined by age, Body Mass Index (BMI), blood pressure, smoking behaviour, diabetes mellitus and physical activity level, which were referenced by Jakarta Cardiovascular Scores (Kusmana, 2002). Data were analysed using the frequency distribution.

## 3 RESULTS

Characteristics of respondents in terms of sex, age, blood pressure, BMI, smoking behaviour, diabetes mellitus and physical activity can be seen in Table 1.

In terms of the characteristics of the respondents, all were women (100%), the most common age range was 45–49 years (18.57%), and the majority (58.57%) had normal blood pressure and a BMI in the range 13.79–25.99 (64.29%), did not smoke (97.14%), did not have diabetes mellitus (100%) and did moderate physical activity (77.14%).

Table 1. Frequency distribution of respondents based on characteristics.

| Characteristic | Frequency | % |
|---|---|---|
| Sex | | |
| Men | 0 | 0 |
| Women | 70 | 100 |
| Age (Year) | | |
| 25–34 | 10 | 14.29 |
| 35–39 | 6 | 8.57 |
| 40–44 | 12 | 17.14 |
| 45–49 | 13 | 18.57 |
| 50–54 | 9 | 12.86 |
| 55–59 | 9 | 12.86 |
| 60–64 | 11 | 15.71 |
| Blood pressure | | |
| Normal | 41 | 58.57 |
| High-Normal | 10 | 14.29 |
| Grade 1 Hypertension | 14 | 20.00 |
| Grade 2 Hypertension | 4 | 5.71 |
| Grade 3 Hypertension | 1 | 1.43 |
| Body mass index | | |
| 13.79–25.99 | 45 | 64.29 |
| 26.00–29.99 | 16 | 22.86 |
| 30.00–35.58 | 9 | 12.86 |
| Smoking | | |
| Never | 68 | 97.14 |
| Ex-smoker | 0 | 0 |
| Yes | 2 | 2.86 |
| Diabetes mellitus | | |
| Normal | 70 | 100 |
| Yes | 0 | 0 |
| Physical activity | | |
| None | 3 | 4.29 |
| Low | 6 | 8.57 |
| Medium | 54 | 77.14 |
| High | 7 | 1.00 |

Table 2. Risk of cardiovascular disease of respondents.

| Risk of cardiovascular disease | Frequency | % |
|---|---|---|
| Low | 37 | 12.86 |
| Moderate | 24 | 34.29 |
| High | 9 | 52.86 |
| Total | 70 | 100 |

Table 3. Knowledge of respondents about coronary heart disease.

| Knowledge | Frequency | % |
|---|---|---|
| Good | 27 | 38.57 |
| Sufficient | 20 | 28.57 |
| Poor | 23 | 32.86 |
| Total | 70 | 100 |

The risk of cardiovascular disease of the respondents can be seen in Table 2. It was found that the majority of respondents (87.15%) had moderate or high-risk factors for cardiovascular disease.

The knowledge of the respondents about coronary heart disease is shown in Table 3. It was found that more than half of the respondents (61.43%) did not have good knowledge of coronary heart disease.

## 4 DISCUSSION

All of the respondents were women. Women have a lower risk of coronary heart disease than men but their risk increases after menopause (Galbut & Davidson, 2005). This is because younger women are protected by the hormone oestrogen, which dilates blood vessels. After the menopause, oestrogen levels in the body decrease, resulting in an increased risk of coronary heart disease. Research shows that as much as 42% of atherosclerotic events are affected by age.The median age of atherosclerosis incidence is 28 years (Main, 2007, cited in Islamee, 2008). The risk of cardiovascular disease increases with age, which increases the morbidity and mortality of this disease at the age of 30–44 years (Anwar, 2004, cited in Islamee, 2008).

The results showed that 41.43% of the respondents had either high-normal blood pressure or hypertension (grades 1, 2 or 3). Research conducted by Nababan (2008) shows that there is a significant relationship between coronary heart disease events and hypertension. Increased blood pressure causes increased arterial wall pressure, which accelerates atherosclerosis and arteriosclerosis thus accelerating the occurrence of rupture and vascular occlusion compared with people who have normal blood pressure (Stern, 1979; Lipoeta, 2006).

Based on BMI, the results showed that 35.71% of respondents were either overweight or obese. Other studies indicate that the risk of coronary heart disease is greater in those classified as overweight when compared with groups of ideal or underweight BMI (Mawi, 2003). Different studies show that there is no significant association of incidence of coronary heart disease with obesity. However, there is an increase in the proportion of patients with obesity and coronary heart disease when compared with coronary heart disease in non-obese patients (Nababan, 2008).

The results showed that there were two (2.86%) respondents who were smokers. In this study, all respondents were women. Supriyono (2008) mentions that women who smoke will experience earlier menopause compared with non-smokers.

The results showed that none of the respondents had diabetes mellitus. A person with diabetes mellitus is at greater risk (200%) of cardiovascular disease than those without diabetes mellitus. A study showed that there was a significant relationship between long-suffering diabetes mellitus with coronary heart disease events in patients with type 2 diabetes mellitus (Yuliani, Oenzil, Iryani, 2014).

The results showed that most respondents, 77.14%, undertook moderate physical activity. Physical activity can reduce the risk of coronary heart disease by increasing High-Density Lipoprotein (HDL) cholesterol levels and coronary collateral repair, and improving lung function and oxygen consumption by the myocardium (Anwar, 2004, cited in Islamee, 2008).

Based on the risk of cardiovascular disease, a majority of respondents (87.15%) had moderate- or high-risk factors for cardiovascular disease. A person with moderate-risk factors has a 10–20% likelihood of developing cardiovascular disease, and a person with high-risk factors has a likelihood of developing cardiovascular disease of more than 20% (Kusmana, 2002).

In terms of knowledge about coronary heart disease, most of the respondents (61.43%) did not have good knowledge.

The results of this study indicate that a majority of the respondents lacked good knowledge of, and were at moderate to high risk of, cardiovascular disease. This is likely to increase the incidence of cardiovascular disease and there is a clear need for primary and secondary prevention efforts such as changing lifestyle and reducing risk factors. This should be supported with adequate knowledge of coronary heart disease. There is a significant relationship between knowledge and the ability to avoid or mitigate coronary heart disease risk factors (Indrawati, 2014).

## REFERENCES


Dalusung-Angosta, A.N. (2010). *Coronary heart disease knowledge and risk factors among Filipino-Americans connected to primary care services* (Doctoral dissertation, University of Hawaii, Manoa, Honolulu, HI). Retrieved from https://scholarspace.manoa.hawaii.edu/bitstream/10125/22042/2/uhm_phd_angosta-a_r.pdf.

Galbut, B.H. & Davidson, M.H. (2005). Cardiovascular disease: Practical applications of the NCEP ATP III update. *Patient Care, March*, 31–38.

Indonesian Ministry of Health. (2011). Guidelines for controlling risk factors for heart disease and blood vessels(1st ed.). Jakarta, Indonesia: Kementerian Kesehatan Republik Indonesia. Retrieved from http://www.pptm.depkes.go.id/cms/frontend/ebook/Pedoman_PJPD_2013.pdf.

Indrawati, L. (2014). Relationship between knowledge, attitude, perception, motivation, family support and informational sources of coronary heart disease with secondary precautions risk factors. *Jurnal Ilmiah Widya, 2*(3), 30–36.

Islamee, A.U. (2008). Cardiovascular risk factors associated with the presence or absence of abnormalities in electrocardiographic results in the Jamaat Assembly Dzikir SBY Nurussalam (Thesis, University of Indonesia).

Kusmana, D. (2002). The influence of smoking cessation, regular physical exercise and/or physical activity on survival: A 13 years cohort study of the Indonesian population in Jakarta. *Medical Journal of Indonesia, 11*(4), 230–241.

Lipoeta, I. (2006). Nutrition and food on cardiovascular disease. Yogyakarta, Indonesia: Andalah Insist Press.

Mawi, M. (2003). Body mass index as determinant of coronary heart disease in adults more than 35 years of age. *Jurnal Kedokteran Trisakti, 23*(3), 87–92.

Nababan, D. (2008). Relationship of risk factors and characteristics of patients with the incidence of coronary heart disesase at Dr. Pirngadi Medan (Thesis, University of North Sumatra, Medan, Indonesia).

Smeltzer, S.C.O., Bare, B.G., Hinkle, J.L. & Cheever, K.H. (2010). *Brunner and Suddarth's textbook of medical-surgical nursing* (12th ed.). Philadelphia, PA: Wolters Kluwer Health.

Stern, M.P. (1979). The recent decline in ischemic heart disease mortality. *Annals of Internal Medicine, 91*(4), 630–640 .

Supriyono, M. (2008). Risk factors for coronary heart disesase in group of aged ≤ 45 year (case study in Dr. Kariadi Hospital Semarang and Telogorejo Hospital Semarang) (Thesis, Diponegoro University, Semarang, Indonesia).

Yuliani, Oenzil, Iryani. (2014). Relationship of various risk factors against coronary heart disease in type 2 diabetes mellitus patients. Jurnal Kesehatan Andalas, 3 (1).

*Strengthening Research Capacity and Disseminating New Findings in Nursing and Public Health – Malini et al. (Eds)*


# The effect of diaphragmatic breathing exercise on pulmonary ventilation function in patients with asthma: A preliminary study

E. Afriyanti & B.P. Wenny
*Faculty of Nursing, University of Andalas, Padang, West Sumatra, Indonesia*

ABSTRACT: Asthma is a chronic inflammatory disease of the respiratory tract, caused by the sensitivity of the trachea and its branching. This study aims to identify the effect of diaphragmatic breathing exercise on pulmonary ventilation function in patients with asthma using the parameter of Peak Expiratory Flow (PEF). This research uses a quasi-experimental method with pretest/post-test and control group design. The selected sample consisted of 20 subjects using random sampling, with ten subjects in the treatment group, and ten control subjects. Data analysis used a *t*-test with $p < 0.05$. The results showed that the pretest PEF was 63.2% and it increased post-test to 90.8% in the treatment group. The PEF also increased in the control group, where the PEF pretest was 63.3% and 86.5% in the post-test. This study showed that diaphragmatic breathing exercise has a significant effect on the increase of PEF in asthma patients.

## 1 INTRODUCTION

Asthma is a serious health problem that affects 300 million people of all ages around the world (Prem et al., 2012). Asthma is a chronic inflammatory disease of the respiratory tract that is reversible, characterised by a wide airway narrowing to varying degrees, resulting in an increased response of the trachea and bronchi to various stimuli, with clinical manifestations of cough, chest tightness due to airway obstruction, as well as episodic wheezing (Henneberger et al., 2011). Currently, asthma still shows a high prevalence. Based on worldwide data from the World Health Organization (WHO, 2010), it is estimated there are 300 million people with asthma and the number will grow to 400 million by 2025. According to the results in the Indonesian Ministry of Health (2013) report, the prevalence of asthma in Indonesia was 4.5%, which underwent a 1% increase from the level reported by Indonesian Ministry of Health (2007), while the ranking of asthma incidence in West Sumatra grew from 2.0% (2007) to 2.7% (2013). Asthma is a disease of airway obstruction with symptoms of cough, wheezing, and shortness of breath. Airway constriction in asthma occurs as a result of bronchial obstruction and spasm of smooth muscle in the bronchus, so that the patient has difficulty in breathing. Expiration is always more difficult and longer than inspiration (Henneberger et al., 2011). Inflammation, mucosal membrane oedema, and mucous hypersecretion in the airways cause difficulty for air to pass through (Price & Wilson, 2006).

People with asthma will tend to breathe in high lung volumes. Average asthma sufferers breathe 3–5 times faster than normal. This condition requires hard work by the respiratory muscles, such that asthma sufferers will have breathing difficulty (Price & Wilson, 2006). Signs that can be found during asthma attacks are the use of additional muscles for breathing (sternocleidomastoid and scalene muscles in the neck). Patients will more often use the respiratory muscles of the chest, rather than the abdominal respiratory muscles, so that the diaphragm muscles cannot relax perfectly. The use of chest-breathing muscles and excessive contraction of the diaphragm muscles can affect the ability of the respiratory muscles as a whole, so that the muscles' capacity slowly decreases and the disease gets worse. This condition causes asthma sufferers to be more frequently and repeatedly treated for increasingly severe complaints, and medically diagnosed with a heavier degree of asthma.

Nurses can perform respiratory therapy to overcome these problems. Respiratory therapy aims to train correct breathing, to flex and strengthen the respiratory muscles, train effective expectoration, and improve circulation.

According to Prem et al. (2012), respiratory therapy in people with asthma is through such means as the Buteyko method, *pranayana* breathing, and diaphragmatic breathing. Diaphragmatic Breathing Exercise (DBE) is performed by maximising the function of the lower lungs so as to increase the capacity of the lungs in breathing, by raising the stomach forward slowly when exhaling (Widarti, 2011). DBE is a respiratory therapy that can increase expiratory air in asthmatics (Ariestianti et al., 2014; Oni et al., 2014), which in turn improves quality of life (Prem et al., 2012).

## 2 METHODS

This research uses a quasi-experimental method with a control group approach and pretest/post-test design. Both control and treatment groups underwent pulmonary ventilation measurements in the form of Peak Expiratory Flow (PEF) measurement (pretest). The treatment group was educated in the use of DBE technique, while the control group was not given the breathing treatment. After the researcher was convinced that the subjects were able to perform DBE correctly, they were required to perform the exercise four times for two minutes every day for one week, in the morning, at noon, in the afternoon, and in the late evening. After the seventh day, post-test PEF measurements were performed.

The entire population of this study consisted of patients with asthma, with a sample size of ten people for each group. The sampling technique used in the research was probability sampling in a simple random sampling. Inclusion criteria were mild and moderate asthma with PEF in the range 60–80%. Participants had to express their willingness to be a research subject, be aged between 18 and 65 years, and not be in exacerbation (PEF $< 60\%$). Both groups received medical drug therapy, which consisted of bronchodilators Ventolin and Aminofilin (aminophylline) infusion at 10 cc 8 hours/500 ml. The tool used to measure pulmonary ventilation function with PEF is a peak flow meter. This research has obtained consent from the Research Ethics Commission at the Faculty of Medicine, University of Andalas. Analysis is made using a paired *t*-test and an independent *t*-test with 95% degree of confidence. Data normality was previously tested and proven normal.

## 3 RESULTS

As shown in Table 1, the distribution of the research subjects is mostly female (80%), in middle adulthood (85%), and has a non-smoking majority (80%).

The results are described in Table 2, which shows that there is a significant difference in the average value of PEF between the treatment and control groups. In the treatment group, the average value of PEF during pretest was 63.2% and went up at the post-test level to 90.9% with $p < 0.05$. Meanwhile, the average value of PEF in the control group increased from 63.3% in the pretest to 86.8% in the post-test with $p < 0.05$.

The purpose of the independent sample t-test in this study was to determine the significance of any difference in PEF values between the treatment group and the control group following administration of the DBE respiratory exercise. The analysis results are shown in Table 3.

Based on Table 3, it can be seen that the average increase in PEF in the treatment group after DBE was 27.5%, with a lower 95% Confidence Interval (CI) value of 3.54 and an upper value of 4.95. In the control group, the average increase in PEF was 23.1%, with a lower 95% CI value of 3.54 and an upper value of 4.96. The result of an independent *t*-test sample in both groups indicated $p < 0.05$, which means that DBE has an influence on the increase of peak expiratory flow in asthma patients.

Table 1. Frequency distribution of research subjects of asthma patients.

| Variable | Control *f* | Control % | Treatment *f* | Treatment % | Total | % |
|---|---|---|---|---|---|---|
| Age | | | | | | |
| Young adult (18–35 years old) | 0 | 0 | 1 | 10 | 1 | 5 |
| Middle adult (36–60 years old) | 10 | 100 | 7 | 70 | 17 | 85 |
| Elderly (> 60 years) | 0 | 0 | 2 | 20 | 2 | 10 |
| Gender | | | | | | |
| Male | 2 | 20 | 2 | 20 | 4 | 20 |
| Female | 8 | 80 | 8 | 80 | 16 | 80 |
| Smoking history | | | | | | |
| Yes | 2 | 20 | 2 | 20 | 10 | 50 |
| No | 8 | 80 | 8 | 80 | 10 | 50 |

Table 2. The mean difference of PEF subjects in the treatment and control groups before and after diaphragmatic breathing exercise.

| Variable | Mean (%) | Min–Max | SD | *p* |
|---|---|---|---|---|
| Treatment group | | | | |
| Pretest PEF | 63.2 | 60.6–68.6 | 2.48 | 0.000 |
| Post-test PEF | 90.8 | 89.0–91.9 | 0.76 | |
| Control group | | | | |
| Pretest PEF | 63.3 | 61.6–65.3 | 1.24 | 0.000 |
| Post-test PEF | 86.5 | 85.4–87.9 | 0.74 | |

Table 3. Differences between the treatment and control groups in PEF values after administration of diaphragmatic breathing exercise.

| Variable | Mean (%) | SD | 95% CI (lower–upper) | *t* | *p* |
|---|---|---|---|---|---|
| Treatment group | 27.5 | 0.7613 | 3.54–4.95 | 12.5 | 0.000 |
| Control group | 23.1 | 0.7499 | 3.54–4.96 | 12.5 | |

## 4 DISCUSSION

Based on Table 2, it can be seen that the average value of the PEF in the treatment group after administration of DBE has increased from an average pretest value of 63.2% to 90.9% in the post-test. From the analysis, the researcher concluded that the highest post-test PEF scores in the treatment group were achieved at the ages of 36 (91.8%) and 42 (91.9%) years old. According to Antoro (2015), age is the most significant aspect affecting the PEF. Theoretically, muscle strength and respiratory function decrease with age. Changes in the respiratory structure begin early in middle adulthood (Guyton & Hall, 2014). An increase in the lowest PEF values was obtained at age 60 years with a value of 89.6%; this is due to the ageing process, which causes decreased alveoli elasticity, a thickening of the bronchial gland, decreased lung capacity, and an increased amount of loss space (Guyton & Hall, 2014). The decrease in lung capacity is caused by the weakening of the intercostal muscles, thereby reducing movement of the chest wall, and the presence of vertebral osteoporosis, thereby decreasing spinal flexibility and enhancing the degree of kyphosis (Antoro, 2015). This also increases further the anteroposterior diameter of the chest cavity. The diaphragm is commonly flatter and loses its elasticity.

Furthermore, the results of the analysis in this study showed that the male subjects did not show a high increase in PEF value, where the highest value of PEF in the treatment group was 89.8% and the lowest was 89.0%. For males in the control group, the highest PEF value was 86.7% and 85.6% the lowest. This is due to the male subjects' smoking history. Cigarettes can adversely affect both active and passive smokers. A person who inhales cigarette smoke on a prolonged basis can suffer from decreased lung function.

Table 3 indicates that the control group subjects also showed an increase in the average value of the PEF without being given a diaphragmatic breathing exercise. Their average pre-test value of PEF was 63.4% and increased to 86.5% post-test. This increase was because the subjects in the control group also received a bronchodilator drug. High PEF values were achieved by 36-year-old (87.9%) and 42-year-old subjects (87.2%). This condition was caused by changes in breathing structure that began early in middle adulthood, at which age decreasing elasticity of the chest wall begins to occur (Guyton & Hall, 2014).

DBE affects the quality of life of subjects with asthma. It was found that DBE can improve asthma patients' quality of life (Prem et al., 2012). Antoro (2015) also discovered that DBE could increase the PEF of people with asthma. Diaphragmatic breathing optimises abdominal movement and restricts chest movement, so that the abdominal muscles here play an important role in the process of expiration and the breathing exercise may influence the increase of work of the abdominal muscles (Ariestianti et al., 2014).

Based on the results of the research, it was found that both the treatment and control groups experienced an increase in PEF value, but the increase was higher in the treatment group. This is in line with the findings of Salvi et al. (2014), which indicate that DBE has an influence on the increased value of spirometry parameters in asthma patients when the exercise is done regularly for one week. This effect occurs because DBE trains the major muscles of respiration such as the diaphragm muscles to work during inspiration, and abdominal muscles to work during expiration. At the time the respiratory process occurs, the respiratory muscles are the most important component of the respiratory pump and they should work well to produce more effective ventilation (Ariestianti et al., 2014).

According to some experts, DBE aims to train patients to use the diaphragm properly and relax the accessory (respiratory) muscles. The exercise is also aimed at increasing the volume of breath flow, reducing functional residue, improving ventilation, and mobilising mucus secretion during postural drainage (Sharma, 2008). With DBE, the thorax and lung cavities develop during inspiration, and expiratory muscles (abdominal muscles) actively contract to facilitate the expulsion of air ($CO_2$) from the thorax cavity. This will lead to increased ventilation, resulting in improved alveoli performance so that gas exchange becomes more effective (Antoro, 2015).

## 5 CONCLUSION

This research has concluded that Diaphragmatic Breathing Exercise (DBE) can increase Peak Expiratory Flow (PEF) in asthma patients. It is suggested that DBE could assist patients in using their diaphragm properly and in relaxing their accessory muscles.

## REFERENCES

Alaparthi, G.K., Augustine, A.J., Anand, R. & Mahale, A. (2016). Comparison of diaphragmatic breathing exercise, volume and flow incentive spirometry, on diaphragm excursion and pulmonary function in patients undergoing laparoscopic surgery: A randomized controlled trial. *Minimally Invasive Surgery*, *2016*, 1–12.

Antoro, B. (2015). Pengaruh Senam Asma Terstruktur Terhadap Peningkatan Arus Puncak Ekspirasi (APE) Pada Pasien Asma. *Jurnal Kesehatan*, *6*(1), 69–74.

Ariestianti, I., Pangkahila, J.A. & Purnawati, S. (2014). Pemberian diaphragmatic breathing sama baik dengan pursed lip breathing dalam meningkatkan arus puncak ekspirasi pada perokok aktif anggota Club Motor Yamaha Vixion Bali di Denpasar [Giving diaphragmatic breathing is equally as good as

pursed lip breathing in improving peak expiratory flow among active smoking members of Yamaha Motorcycle Vixion Club in Denpasar, Bali]. *Majalah Ilmiah Fisioterapi Indonesia, 1*(1). Retrieved from https://ojs.unud.ac.id/index.php/mifi/article/view/8473.
Banjarnahor, P. (2004). *Hubungan antara arus puncak ekspirasi penderita rinosinusitis dengan faktor rhinitis alergi dan tanpa faktor rhinitis alergi [The correlation of peak expiratory flow in rhinosinusitis patients with allergic rhinitis factor and those without allergic rhinitis factor*]. Research report, Medical Faculty, Diponegoro University, Semarang, Indonesia. Retrieved from http://eprints.undip.ac.id/14883/1/2004FK619.pdf.
Enright, S.J., Unnithan, V.B., Heward, C., Withnall, L. & Davies, D.H. (2006). Effect of high-intensity inspiratory muscle training on lung volumes, diaphragm thickness, and exercise capacity in subjects who are healthy. *Physical Therapy, 86*(3), 345–354.
Guyton, A.C. & Hall, J.E. (2014). *Mekanisme fisiologi dan penyakit manusia.* Jakarta, Indonesia: Penerbit Buku Kedokteran.
Henneberger, P.K., Redlich, C.A., Callahan, D.B., Harber, P., Lemiere, C., Martin, J., ... Torén, K. (2011). An official American Thoracic Society statement: Work-exacerbated asthma. *American Journal of Respiratory and Critical Care Medicine, 184*(3), 368–378.
Indonesian Ministry of Health. (2007). *Riset kesehatan dasar.* Jakarta, Indonesia: Kementerian Kesehatan Republik Indonesia.
Indonesian Ministry of Health. (2013). *Riset kesehatan dasar*. Jakarta, Indonesia: Kementerian Kesehatan Republik Indonesia.
Jones, A.Y., Dean, E. & Chow, C.C. (2003). Comparison of the oxygen cost of breathing exercises and spontaneous breathing in patients with stable chronic obstructive pulmonary disease. *Physical Therapy, 83*(5), 424–431.
Lee, H.Y., Cheon, S.H. & Yong, M.S. (2017). Effect of diaphragm breathing exercise applied on the basis of overload principle. *Journal of Physical Therapy Science, 29*(6), 1054–1056.
Mayuni, A.A.I.D., Kamayani, M.O.A. & Puspita, L.M. (2015). Pengaruh diafragmamatic breathing exercise terhadap kapasitas vital paru pada pasien asma di wilayah kerja Puskesmas III Denpasar Utara. *COPING (Community of Publishing in Nursing), 3*(3), 31–36.
Oni, A.O., Erhabor, G.E. & Oluboyo, P.O. (2014). Interchanging spirometric and peak flow meter readings in obstructive airway diseases. *African Journal of Respiratory Medicine, 10*(1), 15–17.
Prem, V., Sahoo, R.C. & Adhikari, P. (2012). Effect of diaphragmatic breathing exercise on quality of life in subjects with asthma: A systematic review. *Physiotherapy Theory and Practice, 29*(4), 271–277. doi:10.3109/09593985.2012.731626.
Price, S.A. & Wilson, L.M. (2006). *Pathophysiology: Clinical concepts of disease processes.* Jakarta, Indonesia: EGC.
Salvi, D., Agarwal, R., Salvi, S., Barthwal, B.M. & Khandagale, S. (2014). Effect of diaphragmatic breathing exercise on spirometric parameters in asthma patients and normal individuals. *Indian Journal of Physiotherapy & Occupational Therapy, 8*(3), 43–48.
Sharma, P.S. (2008, December). *Diaphragmatic and pursed lips breathing.* Mind Publications. Retrieved from http://www.mindpub.com/art574.htm.
Smeltzer, S.C. & Bare, B.G. (2002). *Buku ajar keperawatan medikal bedah Brunner dan Suddarth* (8th ed.). Jakarta, Indonesia: EGC.
World Health Organization. (2010). *World Health Statistic.* Retrieved from http://www.who.int/gho/publications/world_health_statistics/EN_WHS10_Full.pdf.
Widarti. (2011). *Pengaruh diafragmatik breathing exercise terhadap peningkatan mutu hidup pasien asma* (Doctoral dissertation, Muhammadiyah University of Surakarta, Indonesia).
Yunus, F. (2005). *Senam Asma Indonesia.* Jakarta, Indonesia: Yayasan Asma Indonesia, FKUI.

*Strengthening Research Capacity and Disseminating New Findings in Nursing and Public Health – Malini et al. (Eds)*
*© 2018 Taylor & Francis Group, London, ISBN 978-1-138-50066-2*

# Risk factors associated with hypertension: A case study in Andalas Public Health Centre

S. Rahmadani, E. Huriani & R. Refnandes
*Faculty of Nursing, Andalas University, Padang, Indonesia*

ABSTRACT: This study aimed to examine the relationship between family history of hypertension, obesity, exercise habits, smoking duration, and emotional and mental disorders, with hypertension in Andalas Public Health Centre, Padang, Indonesia. This was an analytical study with cross-sectional approach. The sample in this study was 270 people and data was analysed using the chi-squared test and regressions. Hypertension incidence was 33.7%. There were significant relationships between family history, obesity, exercise habits, duration of smoking, and mental and emotional disorders, with the incidence of hypertension ($p < 0.05$). Factors associated most with increased blood pressure were obesity, family history of hypertension, and exercise habits. This study revealed that lifestyle factors correlate with hypertension.

*Keywords*: factors, family history, hypertension, obesity

## 1 INTRODUCTION

It is estimated that 1 billion people worldwide have high blood pressure. Hypertension is one of the main causes of premature death worldwide. In 2020, around 1.56 billion adults are predicted to be living with hypertension. Hypertension kills nearly eight billion people every year in the world and nearly 1.5 million people annually in South-East Asia. A third of adults in South-East Asia suffer from hypertension (WHO, 2015). According to the Basic Health Research (Ministry of Health, 2013) in 2007, hypertension was ranked first in the list of non-communicable diseases in Indonesia, with a prevalence of 31.7% in 2013, while it was 22.6% in West Sumatra.

Some factors that increase morbidity and mortality in the world include being hypertensive, non-availability of water, poor sanitation, child malnutrition and over-nutrition, anaemia in women, diabetes, obesity, alcohol consumption, and smoking. Adults who suffer from hypertension, and who are overweight and obese, may have a higher risk factor for cardiovascular disease and some types of cancer (WHO, 2015).

The prevalence of hypertension continues to rise every year. Besides its effect on the survival of humans and labour productivity, hypertension also adds to the burden of health care costs. Hypertension is a leading cause of stroke, heart disease, and kidney failure. The prognosis is good if an abnormality is detected in the early phase and governance is started before complications occur. A severe increase in blood pressure (hypertensive crisis) can be fatal (Robinson, 2014).

It is important to know the risk factors for hypertension before the complications occur that can take a life. Recognising the risk factors is the first step of appropriate management. For patients suffering from grade 1 hypertension with no other risk factors, the healthy lifestyle strategy is the early-stage treatment. If there have been other complications, it is advisable to perform pharmacological therapy (Soernata et al., 2015).

The 2015 Annual Report of Padang City Health Office showed that hypertension is the disease associated with the highest number of patients (33,647) visiting Padang City Health Centre, as recorded by the National Health Insurance. The Monthly Surveillance Report of Non-Communicable Disease from the public health centre in Padang also recorded hypertension in the top position.

Table 1. Prevalence of hypertension with risk factors.

| Variable | Hypertension % | Unhypertension % | *p* value |
|---|---|---|---|
| Family history | | | |
| Yes | 50.7 | 49.3 | 0.01 |
| No | 27.9 | 72.1 | |
| Obesity | | | |
| Yes | 55.3 | 44.7 | 0.00 |
| No | 20.4 | 79.6 | |
| Exercise habits | | | |
| Ideal | 39.7 | 60.3 | 0.00 |
| Not ideal | 13.1 | 86.9 | |
| Length of time smoking | | | |
| ≥20 years | 38.3 | 61.7 | 0.24 |
| <20 years | 30.7 | 69.3 | |
| Mental emotional disorder | | | |
| Yes | 28.6 | 71.4 | 1.00 |
| No | 33.8 | 66.2 | |

This study aims to examine the relationship between a history of hypertension, obesity, exercise habits, smoking duration, and emotional and mental disorders with hypertension in Andalas Public Health Centre, Padang, Indonesia.

## 2 METHOD

Research design was analytic with cross-sectional approach. This research was conducted. Respondents of this study were people visiting Andalas Public Health Centre, Padang with inclusion criteria of 25–59 years of age. The number of respondents was 270 people. Ethical approval of this study was obtained from relevant institutional ethical committees. Instruments employed in data collection were exercise habit, which was adopted from Rinawati (2012), and the Self Report Questionnaire (SRQ 20) for mental and emotional disorder (WHO, 2013). Data was analysed using chi-squared test and logistic regressions.

## 3 RESULTS

Socio-demographic characteristics of respondents indicated that 32.9% of respondents were 36–45 years old, 54.8% were male, 39.6% passed high school, and 25.2% were self-employed. Furthermore, 33.7% of respondents had hypertension, 25.6% of respondents had a family history of hypertension, 38.1% of the respondents were obese, 77.4% had poor exercise habits, 39.6% of respondents had smoking habit ≥ 20 years, and 2.6% of respondents experienced mental and emotional disorders.

Table 1 indicates that variables having relationships with the incidence of hypertension constitute a family history of hypertension, obesity, and exercise habits. Multivariate analysis was done by multiple logistic regression. Results showed there were still significant relationships when variables of family history, obesity, and exercise habits were analysed together. When they are put in the order of strength of the relationship, obesity (Odds Ratio/OR = 4.154) was the largest, followed by history of hypertension (OR = 1.986), and exercise habits (OR = 0.282). Based on the analysis, the variables are mostly associated with obesity.

## 4 DISCUSSION

The high incidence of hypertension in this study needs attention and precautions so that it does not lead to complications and death (WHO, 2015). This can be done by controlling risk factors of hypertension, especially those factors that have been proven statistically. Because patients with hypertension often have no symptoms, the US National Heart, Lung, and Blood Institute estimated that half of the population who suffer from hypertension are unaware of their condition (National Institutes of Health, 2004).

Besides changes in lifestyle, elevated blood pressure also becomes one risk factor. Controlling blood pressure can be done through activities such as sports, weight control, reducing fat consumption, stopping smoking and alcohol consumption, and good stress management. In addition, hypertensive respondents can get the disease due to unmodifiable factors such as gender, age, and family history of hypertension (Rinawati, 2012). Preventive efforts performed by health centres should be supported by society's active participation, through routine controls of blood pressure, obedience to pharmacological treatment, and exercising non-pharmacological actions such as reducing salt intake, avoiding smoking, doing sports activities, and weight control (Soernata et al., 2015).

A study by Rahayu (2012) showed that an obese individual is 8.44 times more at risk of suffering from hypertension as compared to individuals who are not obese. Sugiharto (2007) identified a significant relationship between family history and hypertension. Subjects with a family history of hypertension faced 4.04 times greater risk of hypertension than those with no family history of hypertension.

Individuals who are obese have possibly even 8.44 times greater risk of suffering from hypertension as compared to those who are not obese (Roslina, 2008; Rahayu, 2012). The maximum body mass index of 39.56 kg/m$^2$ is classified into obese II, and 8.1% of the respondents were in this category. Although this is not a large percentage, it needs attention from nurses and health centres. This is influenced by an increase in cardiac output and sympathetic nerve activity in people with excessive body weight (Price & Wilson, 2006).

Sedentary lifestyle was a strong risk factor for the occurrence of death from cardiovascular disease. Aerobic physical activity such as brisk walking, trotting, and swimming have been proven to lower blood pressure. More noticeable decrease in blood pressure of hypertensive patients and moderate physical activity can also lower blood pressure. Therefore, patients with hypertension are advised to perform physical activity for approximately 30 to 60 minutes per day (Sani, 2010). Moderate regular aerobic exercise (such as swimming for 30–45 minutes 3–4 times per week) may more effectively lower blood pressure than vigorous exercise such as running and jogging. Systolic blood pressure usually decreases by 4–8 mmHg. Isometric physical exercise such as weightlifting can increase blood pressure and should be avoided by patients with hypertension (Black & Hawks, 2009).

This result concerning the relationship between smoking and hypertension contradicts the investigation by Setyanda et al. (2015), in which a significant relationship between smoking duration and hypertension ($p = 0.017$) was demonstrated. This means that the longer the habit of smoking a respondent has, the higher the chances of suffering from hypertension the respondent will experience. Smoking impacts will be in effect after 10–20 years. Smoking also has a dose-response effect, which means that the younger someone starts smoking, the more difficult it will be for them to quit smoking, and the longer they will keep the habit. This eventually increases their risk of developing hypertension (Bustan, 2007).

Furthermore, this study discovered no significant relationship between mental emotional disorder and hypertension. Differences between results of the current research, previous studies, and theories may be due to the fact that each study has a different research focus and is not a special study on hypertensive patients. Another possibility is that subjects of the research had effective stress-coping mechanisms such that the stress did not give rise to emotional mental disorders. In a study of mental disorders, emotional states affect the incidence of hypertension in respondents over the age of 15 years, including elderly respondents. The current research used respondents with an age range of 25–59 years, which means that the elderly age group was not included. As much as 6.6% of total disability is experienced by elderly people over the age of 60 years in relation to mental disorders and neurological disorders. Dementia and depression are the most common neuropsychiatric disorders that most elderly people experience.

The results showed that there is a strong relationship between mental emotional disorder and elderliness (especially 65 years and over) (Idaini et al., 2009). Elderly people are at risk of suffering from emotional disorders and the prevalence of mental emotional disorder is more common among elderly people, with an incidence rate of 23.4% (Qonitah & Isfandiari, 2015).

## 5 CONCLUSION

Incidence of hypertension is correlated to a family history of hypertension, obesity, exercise habits, and smoking duration. The dominant risk factors associated with hypertension were family history, obesity, and exercise. A health promotion programme is expected to sustain, improve, and protect the health of patients and their social environment, which is influenced by public policy in improving knowledge and awareness of health-seeking behaviour to control hypertension.

## REFERENCES


Black, J.M. & Hawks, J.H. (2009). *Medical-surgical nursing: Clinical management for positive outcomes* (8th ed.). St. Louis, MO: Elsevier Saunders.

Bustan, M.N. (2007). *Epidemiologi penyakit tidak menular [Epidemiology of non-communicable diseases]*. Jakarta, Indonesia: Rineka Cipta.

Idaini, S., Suhardi, K.A. & Kristanto, A.Y. (2009). Analisis gejala gangguan mental emosional penduduk Indonesia [Analysis of mental disorders emotional symptoms of population medicine Indonesia]. *Majalah Kedokteran Indonesia*, *59*, 473–479.

Indonesian Ministry of Health. (2013). *Basic Health Research (Riset Kesehatan Dasar)*. Jakarta, Indonesia: Kementrian Kesehatan Republik Indonesia.

National Institutes of Health. (2004). *The seventh report of the Joint National Committee on prevention, detection, evaluation, and treatment of high blood pressure*. Bethesda, MD: National Institutes of Health.

Padang City Health Office. (2015). The 2015 Annual Report.

Padang City Health Office. (2015). The Monthly Surveillance Report of Non-Communicable Disease.

Price, A. & Wilson, M. (2006). *Patofisiologi: konsep klinis proses-proses penyakit (Pathophysiology: clinical concepts of disease processe)* (6th ed., Vol. 1). Jakarta, Indonesia: EGC.

Qonitah, N. & Isfandiari, M.A. (2015). Hubungan antara IMT dan kemandirian fisik dengan gangguan mental emosional pada lansia (Relationship between body mass index and physical indepence with mental emotional disturbances among elderly). *Jurnal Berkala Epidemiologi*, *3*(1), 1–11.

Rahayu, H. (2012). *Faktor risiko hipertensi pada masyarakat RW 01 Srengseng Sawah, Kecamatan Jagakarsa Kota Jakarta Selatan (Risk factors for hypertension among population in Srengseng Sawah, district of Jagakarsa, South Jakarta)* (Unpublished undergraduate thesis, University of Indonesia, Jakarta, Indonesia).

Rinawati. (2012). *Kesehatan keluarga* [*Family health*]. Jakarta, Indonesia: Tugu.

Robinson, J.M. (2014). *Buku ajar organ system: Visual nursing kardiovaskuler.* (Handbook ofsystem organ: Jakarta, Indonesia: Bina Rupa Aksara.

Roslina. (2008). *Analisa Determinan Hipertensi Esensial di Wilayah Kerja Tiga Puskesmas Kabupaten Deli Serdang Tahun 2007 (Analysis of Determinats of Primar Hypertension in the Working Area of Three Health Centres in Deli Serdang District year 2007)*. (Master's thesis, University of North Sumatra, Medan, Indonesia). Retrieved from http://repository.usu.ac.id/handle/123456789/6783.

Sani, A. (2010). *Hypertension: Current perspectives*. Jakarta, Indonesia: Medya Crea.

Setyanda, Y.O.G., Sulastri, D. & Lestari, Y. (2015). Hubungan merokok dengan kejadian hipertensi pada laki-laki usia 35–65 tahun di Kota Padang. (Relationship between smoking and hypertension among 35–65 years old male in Padang) *Jurnal Kesehatan Andalas*, *4*(2), 434–439.

Soernata, A.A., Erwinanto, Mumpuni, A.S.S., Barack, R., Lukito, A.A., Hersunarti, N. & Pratikto, R.S. (Eds.). (2015). *Pedoman tatalaksana hipertensi pada penyakit kardiovaskular [Guidelines for therapy and management of hypertension in cardiovascular disease in Indonesia*]. Jakarta, Indonesia: Indonesian Heart Association. Retrieved from http://www.inaheart.org/upload/file/Pedoman_TataLaksna_hipertensi_pada_penyakit_Kardiovaskular_2015.pdf.

Sugiharto, A. (2007). *Faktor-faktor resiko hipertensi grade II pada masyarakat Karanganyar* [*Risk Factors of Grade II Hypertension in Community*] (Master's dissertation, Diponegoro University, Semarang, Indonesia). Retrieved from http://eprints.undip.ac.id/5265/.

WHO. (2013).the Self Report Questionnaire (SRQ 20) for mental and emotional disorder. Geneva: Switzerland: World Health Organization.

WHO. (2015). *World health statistics 2015*. Geneva: Switzerland: World Health Organization. Retrieved from http://www.who.int/gho/publications/world_health_statistics/2015/en/.

*Strengthening Research Capacity and Disseminating New Findings in Nursing and Public Health – Malini et al. (Eds)*
*© 2018 Taylor & Francis Group, London, ISBN 978-1-138-50066-2*

# A retrospective study of the association between the quantity and varieties of fruit consumption with the glycaemic status of patients with type-2 diabetes mellitus

J. Widiyanto, Isnaniar & T.K. Ningrum
*Faculty of Mathematics, Natural Sciences and Health, Muhammadiyah University of Riau, Pekanbaru, Riau, Indonesia*

ABSTRACT: The aim of this study was to analyse the correlation of the supply and diversity of fruit consumption associated with the glycaemic status of patients with type-2 diabetes disease in Puskesmas Sidomulyo and Harapan Raya, Pekanbaru. Portable diabetes equipment was used to measure the diabetes level and a cross-sectional survey was developed to assess the amount and diversity of fruit consumption. The study was conducted using case control study methods, and a chi-squared test using SPSS software. This study showed a significance between the supply and diversity of fruit consumption associated with glycaemic status, with p-values <0.05 of about 0.009 and 0.017, respectively: odds ratio of 3.6 and 95% confidence interval: 1.22–10.61. In conclusion, there was a strong correlation between the supply and diversity of fruit consumption and the glycaemic status of patients with type-2 diabetes disease.

## 1 INTRODUCTION

Non-infectious diseases such as Diabetes Mellitus (DM) have become a threat to society. The rate of diabetes in Indonesia is about 15.2%. This reflects a significant rise in the number of patients with diabetes, from 8,426,000 in 2000 to 21,257,00 in 2030 (PERKENI, 2011). Indonesia is ranked fourth in the world for the number of people suffering from diabetes. Only 50% of diabetics are aware that they have diabetes and only 30% of them are treated regularly (American Diabetes Association, 2006). Based on the report of *Riset Kesehatan Dasar (Riskesdas)* in 2013, the prevalence of diabetes in Indonesia is about 1.5%. The research conducted by DiabCare showed that 47.2% of diabetics have poor control of their fasting plasma glucose: >130 mg/dl in patients with type-2 diabetes mellitus (Soewondo & Pramono, 2011).

Soewondo and Pramono (2011) conducted their research based on *Riskesdas* data and found that 4.1% of people are not diagnosed and only 1.6% of the 5.7% of diabetics are diagnosed. Epidemiological studies show a tendency for an increasing number of patients and prevalence of type-2 diabetes all around the world (Cheng, 2005). The World Health Organization (WHO) predicts that this number will continue to increase in subsequent years and reach 21.3 million people by 2030. As well as WHO, the International Diabetes Federation also predicts an increase in the number of diabetics, from about 7 million people in 2009 to 12 million people in 2030 (Shaw et al., 2010).

Some evidence indicates that diabetes complications can be prevented by controlling the glycaemic status optimally. In Indonesia, this solution cannot yet be sought because of the heterogeneous and lack of awareness (Artanti & Rosdiana, 2015).

In the laboratory, metabolic control can be assessed from the routine glycaemic monitoring at home of HbA1c and fructosamine. Fructosamine is the result of a non-enzymatic Maillard reaction between glucose and amino acid residues (Natah et al., 1997). Fructosamine examination can be used as an alternative to HbA1c examination for glycaemic index control, because it is technically easier and cheaper, and the glycaemic control time is shorter than HbA1c (2–4 weeks versus 6–12 weeks) (Petitti et al., 2001).

Some research on the glycaemic diet index shows contradictory results. On the one hand, it is reported that there is a correlation between glycaemic status and the diet index, but on the other hand it is reported that there is no correlation (Evert et al., 2013). Patients with a low glycaemic diet index have a low HbA1c level of 0.43% (confidence interval 0.72–0.13) and there was a decrease in trigricol protein of 7.4%. Meta-analysis studies show that a high-fibre diet has an effect on blood glucose levels (Anderson et al., 2004).

In Indonesia, from the *Riskesdas* 2013, it was seen that the number of diagnosed DM patients amounted to 41,071 people, with symptoms in 8,214 people, and the Riau Health Service data in 2013showed the highest number of diabetes mellitus patients in the age group 45–54 years (251 cases), then in the age group 60–69 years (130 cases), and then a third age group of 25–44 years (126 cases). Meanwhile, the Riau Health Profile data of 2013stated that in Pekanbaru the volume of diabetes mellitus was 507 cases and the Research Riani at Central Health Community Sidomulyo, District of Tampan, showed from the results of 120 subjects studied that 29.17% had undiagnosed diabetes mellitus. Glycaemic status changes in diabetics are affected by many factors, one of which is the pattern of food consumption, especially fruits, so the quantity and varieties of fruit consumed by the patient greatly determines their glycaemic status (Yusra, 2011). Research has shown that the consumption of varied fruits and vegetables may reduce the risk factors for type-2 diabetes (Cooper et al., 2012).

## 2 METHODS

This research is an observational study and the design is a case control study (Armenian, 2009), a research design that uses measurement or observation by way of determining cases and control then cekt backward variable treatment gradually. The research conducted measurements or observations about the relationship between the quantity and varieties of fruit consumption with the status of glycaemia in patients with type-2 diabetes mellitus.

The variables measured or observed in this study are the quantity of fruit consumed by DM patients, categorised as *enough* and *not enough,* the varieties of fruit consumption are categorised as *varied* and *not varied* and the variable status of glycaemic patients with DM measured glycaemic levels that were categorised as *hyperglycaemia* or *not*. The data analysed in this study was conducted using computer aids and the SPSS statistical software for Windows, release 17.0, with stages of analysis as follows: univariate analysis is done by finding the frequency distribution of the variable. The results of this analysis are presented in the form of tables and narratives that include: characteristics of diabetics, quantity of fruit consumption of diabetics, variety of fruit consumption of diabetics, and glycaemic status of diabetics. Bivariate analysis is conducted to test the relationship between the two variables, the independent variable with the dependent variable. The statistical test used is the Pearson chi-squared test to calculate significance. The level of confidence is determined at $p = 0.05$, with a confidence interval of 95%. If the value of $p > 0.05$ then the research hypothesis is rejected. If

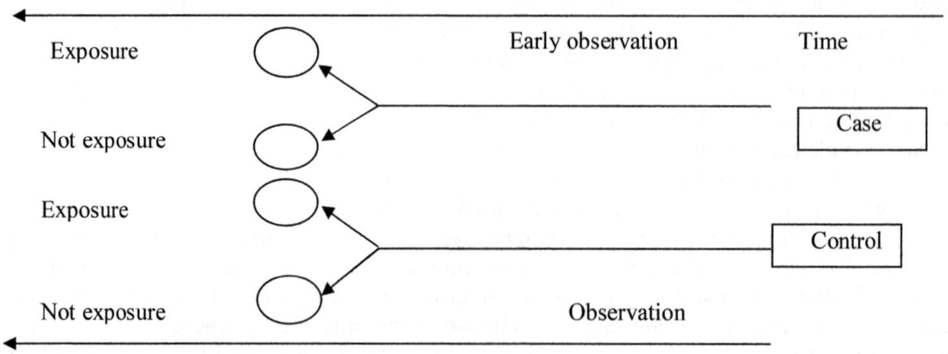


Figure 1. Case control study design.

the value of $p < 0.05$ then the research hypothesis is accepted. Furthermore, it also obtained a large value of risk (odds ratio / OR) exposure to the case by using Table 2 × 2 as follows.

## 3 RESULTS

The results of the research that has been done in Harapan Raya and Sidomulyo Health Centre are illustrated in Table 1.

### 3.1 *Univariate analysis*

Based on Table 1, it can be seen that the percentage of respondents of female gender was 93.9%, the percentage of respondents with a higher level of education was 69.7%, the percentage of respondents who work as a housewife was 72.7%, and 59.1% of respondents were of an age associated with DM risk. Based on the above characteristics, which has correlation glycaemic status is age

Diabetes mellitus is a group of metabolic diseases that have the characteristics of hyperglycaemia, which occurs due to an insulin secretion disorder, insulin work, or both. Metabolic disorders are usually associated with a decrease in the function of a person's organs, especially the function of the pancreas (Rahayu et al., 2013). The results of *Riskesdas* > 15 years of age exposed to a less healthy lifestyle over five years potentially against metabolic and cardiovascular disorders (Okosun & Lyn, 2015).

Diabetes mellitus can attack citizens from various layers of society, both in terms of low, middle and upper economic status, and also in terms of age. Young and old can become diabetes mellitus patients. Generally, humans experience physiological changes that decrease rapidly after the age of 40 years. Diabetes often appears after someone enters a prone age, especially after the age of 45 years in those who are overweight, so that their body is insensitive to insulin. The existing theory says that a person ≥ 45 years of age has an increased risk of diabetes mellitus and glucose intolerance caused by degenerative factors, for example, decreased body function, especially the ability of β cells to produce insulin (Egede & Ellis, 2010)

The prevalence of diabetes cases in the world in the age range 20–79 years is 6.4% of the 285 million adults in 2010 and is expected to increase to 7.7% of 439 million adults in 2030. Between 2010 and 2030 there was an increase of 69% in developed countries and 20% in developing countries (Shaw et al., 2010).

The results of this study are also in line with previous research conducted by Ayu and Indirawati, who argued that the prevalence of diabetes mellitus in Indonesia in the population

Table 1. Distribution of respondent characteristic frequency (n = 66).

| No | Variable | n | % |
|---|---|---|---|
| 1 | Gender: | | |
| | Men | 4 | 6.1 |
| | Women | 62 | 93.9 |
| 2 | Education level: | | |
| | Low | 20 | 30.3 |
| | High | 46 | 69.7 |
| 3 | Work: | | |
| | Government employees | 2 | 3.0 |
| | Private | 1 | 1.5 |
| | Housewife | 48 | 72.7 |
| | Other | 15 | 22.7 |
| 4 | Age: | | |
| | Risk | 39 | 59.1 |
| | Not risk | 27 | 40.9 |

Table 2. Relationship between quantity and varieties of fruit consumption with glycaemic status.

| | Glycaemic status | | | |
|---|---|---|---|---|
| Variable | High | Low | $p$-value | OR/95% CI |
| Varieties | | | | |
| Varieties | 9 | 33 | 0.017 | 3.66/1.23–10.88 |
| Not varieties | 12 | 12 | | |
| Quantity | | | | |
| <3 fruit | 12 | 11 | 0.009 | 0.2/0.081–0.729 |
| ≥3 fruit | 9 | 34 | | |

aged >15 years increased from 1.5–2.3% to 5.6% in 1993 (Muthukumaran et al., 2013). The results of the epidemiological study of diabetes mellitus in Manado also showed a higher figure of 6.1%. The path with what was put forward by Ayu and Indirawati on his study (Betteng & Mayulu, 2014).

### 3.2 *Bivariate analysis*

Based on Table 2, it can be seen that there is a significant relationship between the varieties of fruit consumption with glycaemic status, which is proved by the value of $p < 0.05$, which is 0.017, with an Odds Ratio (OR) of 3.66 and 95% Confidence Interval (CI) of 1.23–10.88.

Also, there is a significant relationship between the quantity of fruit consumption with glycaemic status, which is proved by $p$-value < 0.05, which is 0.009, with an OR of 0.2 and 95% CI of 0.081–0.729.

The variety and quantity of fruit consumption are associated with glycaemic status, although they are not a risk for the incidence of type-2 diabetes mellitus.

The study provides an illustration that the variety and quantity of fruit consumption are closely related to the glycaemic status, and that patients whose quantity of fruit consumption is small have 3.6 times the risk of increased glucose when compared with patients who consume large amounts of fruit.

This is in accordance with the critical review results, which explain that the consumption of vegetables and fruits can improve the prevention of diabetes, stroke and coronary heart disease, including obesity (Boeing et al., 2012).

The study was also supported by the results of a study proving that red dragon fruit juice can reduce the blood glucose levels of male white made rats diabetic at all doses proportional to the hypoglycaemic effects of glibenclamide (Muthukumaran et al., 2013). Research in the US proves that the consumption of foods that are high in carbohydrates and low in fibre has a risk of increased incidence of type-2 diabetes mellitus, so the quantity and quality of high carbohydrate and low fibre consumption is a very important factor in reducing the incidence of diabetes mellitus (Sluijs et al., 2010). The fibre content of fruit could be one factor that could reduce the risk of diabetes mellitus, because the sugar content tends to be more in line with the physiology of the human (Mokhdad et al., 2003).

Nutritional therapy is highly recommended for managing and controlling diabetes mellitus in adults, especially in patients with a history of diabetes mellitus (Alison et al., 2012).

## 4 CONCLUSION

In conclusion, there was a significant correlation between the supply and diversity of fruit consumption and glycaemic status in patients with type-2 diabetes disease. A lack of fruit consumption can cause blood sugar levels that are 3.6 times higher.

## REFERENCES


American Diabetes Association. (2006). Position statement: Standard of medical care in diabetes–2006. *Diabetes Care, 29*(1), S4–S42.

Anderson, J.W., Randles, K.M., Kendall, C.W.C. & Jenkins, D.J.A. (2004). Carbohydrate and fiber recommendations for individuals with diabetes: A quantitative assessment and meta-analysis of the evidence. *Journal of the American College of Nutrition, 35*, 5–17.

Armenian, H.K. (2009). *The case-control method: Design and applications*. New York, NY: Oxford University Press.

Artanti, P.M. & Rosdiana, D. (2015). Angka Kejadian Diabetes Melitus Tidak Terdiagnosis pada Masyarakat Kota Pekanbaru. *Jom FK, 2.*

Betteng, R.D.P. & Mayulu, N. (2014). Analisa Faktor Risiko Penyebab Terjadinya Diabetes Mellitus Tipe 2 Pada Wanita Usia Produktif di Puskesmas Wawonasa. *Jurnal e-Biomedik* (eBM), *2*(2).

Boeing, H., Bechthold, A., Bub, A., Ellinger, S., Haller, D., Kroke, A., ... Watzl, B. (2012). Critical review: Vegetables and fruit in the prevention of chronic diseases. *European Journal of Nutrition, 51*, 637–663.

Cheng, D. (2005). Prevalence, predisposition and prevention of type II diabetes. *Nutrition & Metabolism, 2*(1), 29.

Cooper, A.J., Sharp, S.J., Lentjes, M.A.H., Luben, R.N., Khaw, K.T., Wareham, N.J. & Forouhi, N.G. (2012). A prospective study of the association between quantity and variety of fruit and vegetable intake and incident type 2 diabetes. *Diabetes Care Journal, 35*, 1293–1300.

Egede, L. & Ellis, C. (2010). Diabetus mellitus II and depression. *Global Perspectives, 87*(3), 302–312.

Evert, A.B., Boucher, J.L., Cypress, M., Dunbar, S.A., Franz, M.J., Mayer-Davis, E.J., ... Yancy, W.S., Jr. (2013). Nutrition therapy recommendations for the management of adults with diabetes. *Diabetes Care, 36*(11), 3821–3842.

Mokhdad, A.H., Ford, E.S., Bowman, B., Dietz, W.H., Vinicord, F., Bales, V.S. & Marks, J.S. (2003). Prevalence of obesity, diabetes, and obesity-related high risk factors, 2001. *JAMA, 289*(1), 76–79.

Muthukumaran, J., Srinivasan, S., Venkatesan, R.S., Ramachandran, V. & Muruganathan, U. (2013). Syringic acid, a novel natural phenolic acid, normalizes hyperglycemia with special reference to glycoprotein components in experimental diabetic rats. *Journal of Acute Disease, 2*(4), 304–309.

Natah, S.S., Hussien, K.R., Tuominen, J.A. & Koivisto, V.A. (1997). Metabolic response to lactitol and xylitol in healthy men. *American Journal of Clinical Nutrition, 65*, 947–950.

Okosun, I.S. & Lyn, R. (2015). Prediabetes awareness, healthcare provider's advice, and lifestyle changes in American adults. *International Journal of Diabetes Mellitus, 3*(1), 11–18.

PERKENI. (2011). *Konsensus: Pengelolaan Dan Pencegahan Diabetes Melitus Tipe 2 di Indonesia*. Semarang, Indonesia: Perkumpulan Endokrinologi Indonesia.

Petitti, D.B., Contreras, R. & Dudl, J. (2001). Randomized trial of fructosamine home monitoring in patients with diabetes. *Effective Clinical Practice, 4*, 18–23.

Rahayu, L., Zakir, L. & Keban, S.A. (2013). The effect of rambutan seed (Nephelium lappaceum L.) infusion on blood glucose and pancreas histology of mice induced with Alloxan. *Jurnal Ilmu Kefarmasian Indonesia, 11*(1), 28–35.

Shaw, J.E., Sicree, R.A. & Zimmet, P.Z. (2010). Global estimates of the prevalence of diabetes for 2010 and 2030. *Diabetes Research and Clinical Practice, 8*(7), 4–14.

Sluijs, I., van der Schouw, Y.T., van der A, D.L., Spijkerman, A.M., Hu, F.B., Grobbee, D.E. & Beulens, J.W. (2010). Carbohydrate quantity and quality and risk of type 2 diabetes in the European Prospective Investigation into Cancer and Nutrition–Netherlands (EPIC-NL) study. *American Journal of Clinical Nutrition, 92*, 905–911.

Soewondo, P. & Pramono, L.A. (2011). Prevalence, characteristic, and predictors of pre-diabetes in Indonesia. *Medical Journal of Indonesia, 20*(4), 283–294.

Yusra, A. (2011). *Hubungan Antara Dukungan Keluarga Dengan Kualitas Hidup Pasien Diabetes Militus Tipe 2 Di Poliklinik Penyakit Dalam Rumah Sakit Umum Pusat Fatmawati Jakarta* (Master's thesis, University of Indonesia, Depok, Indonesia).

*Strengthening Research Capacity and Disseminating New Findings in Nursing and Public Health – Malini et al. (Eds)*
 *ISBN 978-1-138-50066-2*

# A dietary assessment of nurses in Klang Valley, Malaysia

M.R. Ab Hamid
*Centre of Nutrition and Dietetics, Faculty of Health Sciences, Universiti Teknologi MARA, Selangor, Malaysia*

N.I. binti Ahamad Sidek
*Universiti Teknologi MARA, Selangor, Malaysia*

N. Said
*Centre of Nursing, Faculty of Health Sciences, Universiti Teknologi MARA, Selangor, Malaysia*

ABSTRACT: This study was carried out in order to determine the dietary intake of nurses at the local hospital in comparison with the recommended nutrient intake for Malaysians. A total of 55 nurses participated in this study. Demographic profiles were taken, and three-day self-administered dietary record forms were given out. The mean daily energy intake based on age group was 1,156 ± 319 kcal for 19–29 year olds and 1,267 ± 422 kcal for 30–50 year olds. Meanwhile, in the group of 19–29 year olds, the distribution of the daily carbohydrate, protein, fat and fibre intake was 126 ± 48 g, 46 ± 28 g, 51 ± 19 g and 1.8 ± 1.3 g, while for 30–50 year olds it was 166 ± 60 g, 47 ± 16 g, 51 ± 22 g and 2.2 ± 1.4 g, respectively. From the results it was shown that the dietary intake among the nurses was below the recommended intake. Therefore, it is suggested that nutritional education, emphasising the importance of a healthy and balanced diet, should be given to the nurses.

*Keywords*: Dietary intake, Dietary Assessment, Nurses

## 1 INTRODUCTION

Nurses, as part of healthcare teams, are necessary to support the healthcare system. They provide patient care and support services to the organisation and to patients. In Malaysia, the majority of nurses in the hospital are working shifts. Working shifts will indirectly influence their dietary pattern, eating habits and nutrient intake, including macronutrients and micronutrients. Previous studies have claimed that shift work is associated with unhealthy and improper eating habits (Lowden et al., 2010; Heath et al., 2012). The chronic intake of an unbalanced diet may compromise their quality of work and impair their nutritional status. A poor quality diet has considerable effects on the economy and health; it is linked to obesity, diabetes, cardiovascular disease, osteoporosis, dental disease and cancer (Beebe et al., 2017).

In Malaysia, the intake levels of essential nutrients based on scientific knowledge that are judged to be adequate to meet the nutrient needs of all healthy individuals are defined as the Recommended Nutrient Intake (RNI). In fact, this recommendation is the key standard by which the nutrients in food consumed are measured as adequate for any given population, particularly in Malaysia (Mirnalini et al., 2008). Many studies have compared the dietary intake of those nurses working during the day and those working shifts, but have not taken into account the recommended nutrient intake. Therefore, when making an assessment of the dietary intake among nurses, it is necessary to know whether or not the nurses are taking in sufficient amounts of the energy and nutrients recommended for the Malaysian population. This study aimed to assess the adequacy of energy, macronutrients and fibre intake among the nurses.

## 2 METHODS

### 2.1 *Study design and sampling technique*

This study is a cross-sectional study conducted in Klang Valley, Malaysia. The subjects were recruited via convenience sampling. The subjects must be working shifts and have no history of diabetes, cardiovascular disease, hypertension or renal failure. However, some subjects who met the inclusion criteria were excluded from this study if they were working part-time jobs, had food intake restrictions due to diseases, participated in any weight management programme or were either pregnant or vegetarian. A total of 55 subjects were eligible and participated in this study on a voluntary basis.

### 2.2 *Ethical considerations*

The consent form was given to the subjects before the data collection. This study received ethical approval from the university (REC/14/15).

### 2.3 *Data collection*

A self-administered questionnaire was used to gather the demographic data of the subjects. A three-day dietary record form was given to the subjects. They were required to record their intake during different working shifts (morning, afternoon and night). A briefing and tutorial regarding completion of the dietary record were given to the subjects by a trained researcher. A sample of food records and pictures of household measurement tools were provided for assistance.

### 2.4 *Data analysis*

The food records received from the subjects were analysed to obtain their energy, macronutrients and fibre intake. The foods items were quantified in grams prior to analysis. The analysis of the nutrients was carried out by referring to the Malaysian Nutrient Composition Table, Food Atlas, Singapore Food Composition, the nutrient label on the packaging and the United States Department of Agriculture Food Composition Databases. These databases provide valid nutrient information about the food content. Subjects were grouped into two age categories: 19–29 and 30–50 years old. This is due to differences in requirements according to the RNI guidelines.

The statistical analysis was conducted using the Statistical Package for Social Sciences (SPSS) software, version 21.0 (SPSS Inc., Chicago, IL, USA). A one-sample *t*-test was employed to compare the dietary intake with the Recommended Nutrient Intake (RNI) for Malaysians (NCCFN, 2005). The level of significance (*p*-value) was set at less than 0.05.

## 3 RESULTS

### 3.1 *Demographic data*

Fifty-five nurses participated in this study. They were grouped into two categories according to their age: 19–29 and 30–50 years old (Table 1). The separation is based on the differences in RNI for the Malaysian population. Most of the subjects (91.0%) were Malays, most were married (63.6%) and 69.1% had been working for more than three years.

### 3.2 *Dietary intake*

The mean daily energy intake of the nurses in the 19–29 year old category, which was 1,156.3 ± 318.9 kcal/day (Table 2), was inadequate ($p < 0.001$) compared to the recommendation. The mean protein intake of the nurses of 46.1 ± 12.8 g/day was also lower than the recommended

Table 1. Demographic data.

| Characteristic | Frequency (n) | Percentage (%) |
|---|---|---|
| Age | | |
| 19–29 | 29 | 52.7 |
| 30–50 | 26 | 47.3 |
| Ethnicity | | |
| Malay | 50 | 91.0 |
| Indian | 4 | 7.3 |
| Others | 1 | 1.8 |
| Marital status | | |
| Single | 18 | 32.7 |
| Married | 35 | 63.6 |
| Widow/Widower | 2 | 3.6 |
| Working experience | | |
| 6 months–1 year | 7 | 12.7 |
| 1–3 years | 10 | 18.2 |
| > 3 years | 38 | 69.1 |

Table 2. Comparison of energy, macronutrients and fibre intake with RNI for women 19–29 years old (n = 29).

| Component | Mean + SD | Macronutrient (%) | RNI (%) | >RNI (%) | RNI* | *p*-value |
|---|---|---|---|---|---|---|
| Energy (kcal/day) | 1,156.3 ± 318.9 | | 29 (100) | 0 (0.0) | 2,000 | 0.00** |
| Carbohydrate (g/d) | 125.5 ± 48.3 | 43.4 | 29 (100) | | 55–75% | |
| Protein (g/d) | 46.1 ± 12.8 | 15.9 | 22 (76) | 7 (24.1) | 55 | 0.00** |
| Fat (g/d) | 51.4 ± 18.8 | 40.0 | 22 (100) | | 20–30% | |
| Dietary fibre (g/d) | 1.8 ± 1.3 | | 22 (100) | | 20–30 | |

*RNI for Malaysians, 2005.
**$p < 0.01$, there is a very significant difference between the energy and nutrient intakes among subjects with RNI using one-sample *t*-test (sig. 2-tailed).

Table 3. Comparison of energy, macronutrients and fibre intake with RNI for women 30–50 years old (n = 26).

| Nutrient | Mean + SD | Macronutrient (%) | RNI (%) | >RNI | RNI* | *p*-value |
|---|---|---|---|---|---|---|
| Energy (kcal/day) | 1,266.9 ± 421.9 | | 25 (96.2) | 1 (3.8) | 2,180 | 0.00** |
| Carbohydrate (g/d) | 165.5 ± 59.6 | 52.3% | | 26 (100) | 55–75% | |
| Protein (g/d) | 46.9 ± 16.3 | 14.8% | 19 (73.1) | 7 (26.9) | 55 | 0.00** |
| Fat (g/d) | 50.6 ± 22.2 | 35.9% | | 26 (100) | 20–30% | |
| Dietary fibre (g/d) | 2.2 ± 1.4 | | 26 (100) | | 20–30 | |

*RNI for Malaysians, 2005.
**$p < 0.01$, there is a very significant difference between the energy and nutrient intakes among subjects with RNI using one-sample *t*-test (sig. 2-tailed).

intake ($p < 0.001$). The carbohydrate intake was below and the fat intake above the respective recommended ranges. The daily fibre intake was insufficient as it was only 1.8 ± 1.3 g/day and well below the recommendation, which is 20–30 g/day.

The mean daily energy intake of the nurses in the 30–50 year old category was 1,266.9 ± 421.9 kcal/ day, which was lower than the recommended intake ($p < 0.001$) (Table 3). The mean protein intake of the nurses of 46.9 ± 16.3 g/day was also lower than the recommended intake ($p < 0.001$). All of the nurses were taking carbohydrates within the recommended range, while the fat intake was higher than RNI. The daily fibre intake was deficient as it was only 2.2 ± 1.4 g/day and did not meet the recommendation, which is 20–30 g/day.

## 4 DISCUSSION

The overall mean energy intake of the nurses for both categories was below the recommended nutrient intake for Malaysians. Nurses in the 30–50 year old age group were taking a higher amount of energy than those in the 19–29 year old age group. The distribution of fat and carbohydrates were either below or above the recommended range. There is a discrepancy in opinion with regards to the effect of working-time arrangements and dietary intake. Previous research has shown that people working night shifts tend to have poorer diets compared to those working during normal daily working hours (Wong et al., 2010; Strzemecka et al., 2014). In contrast, Bonnell et al. (2017) documented that there is no difference in terms of the mean energy intake between nurses working night shifts and those working throughout the day. This could be the reason for the inadequate energy intake among the nurses in this study, as the dietary intake was a combination of three working shifts. The higher energy intake in the group of 30–50 year olds compared to the 19–29 year olds is contradictory to the findings of Mirnalini et al. (2008), who found a reduction in energy intake with age. Another study from Sahu and Dey (2011) found the daily intake of carbohydrate, protein and fat was significantly lower in those working night shifts. On the other hand, the daily energy intake in those working night shifts was lower than for those nurses working a morning, afternoon and off-day in rotating shifts. This difference affected the overall dietary intake of the nurses.

Aside from energy and macronutrients, total fibre intake was lower than the recommendations. The dietary intake of nurses is affected by several factors. Varli and Bilici (2016) reported that one such factor was that nurses working shifts skipped their meals, especially breakfast, and this eventually caused a reduction in their total energy, macronutrient and micronutrient intakes. In addition, Ilmonen et al. (2012) found that the majority of nurses reported an insufficiency in nutrition components in nursing education, causing them to be less aware of the importance of good nutrition (Beebe et al., 2017).

The findings of this research may indicate the level of nutritional knowledge and its practice among the nurses. According to previous studies, training updates on nutritional knowledge for registered nurses are limited (Schaller & James, 2005). The American Dietetic Association (1998) emphasises the importance of nurses having at least a basic knowledge of nutrition. Our study similarly suggests a need for further nutrition training and education for nurses. Health initiatives should also be implemented among both day and night shift nurses (Beebe et al., 2017). In addition, Bonnell et al. (2017) suggested that modification of the workplace environment can support healthier food choices for those working both at night and during the day and influences positive health behaviours. Previous research indicates that nutrition training programmes for nurses can be effective in helping to increase nutritional knowledge and awareness (Bjerrum et al., 2012). Training can focus on consuming nutrient-dense foods, such as fruits, vegetables, whole grains and lean meats, while simultaneously limiting the intake of saturated and trans fats, added sugars and cholesterol to improve the diet quality of nurses working both day and night shifts (Beebe et al., 2017).

## 5 CONCLUSION

In conclusion, the mean energy, protein and fibre intake of the nurses was below the recommended levels. The majority of the carbohydrate and fat intakes were either below or above the recommended range. Hence, nutrition education programmes should be implemented in order to improve the knowledge, attitude and practice among nurses.

## REFERENCES


American Dietetic Association. (1998). Position of the American Dietetic Association: Nutrition education for health care professionals. *Journal of the American Dietetic Association, 98*(3), 343–346.

Beebe, S.D.M., Jen Chang, D.J.P., Kress, S.K.M., Mattfeldt-Beman, D.M.P. & Debi Beebe, C. (2017). Diet quality and sleep quality among day and night shift nurses. *Journal of Nursing Management, 25*(7), 549–557.

Bjerrum, M., Tewes, M. & Pedersen, P. (2012). Nurses' self-reported knowledge about and attitude to nutrition: Before and after a training programme. *Scandinavian Journal of Caring Sciences, 26*(1), 81–89.

Bonnell, E.K., Huggins, C.E., Huggins, C.T., McCaffrey, T.A., Palermo, C. & Bonham, M.P. (2017). Influences on dietary choices during day versus night shift in shift workers: A mixed methods study. *Nutrients, 9*(3), 1–13.

Heath, G., Roach, G.D., Dorrian, J., Ferguson, S.A., Darwent, D. & Sargent, C. (2012). The effect of sleep restriction on snacking behaviour during a week of simulated shift work. *Accident Analysis and Prevention, 45*(Suppl), 62–67.

Ilmonen, J., Isolauri, E. & Laitinen, K. (2012). Nutrition education and counselling practices in mother and child health clinics: Study amongst nurses. *Journal of Clinical Nursing, 21*(19/20), 2985–2994.

Lowden, A., Moreno, C., Holmback, U., Lennernas, M. & Tucker, P. (2010). Eating and shift work: Effects on habits, metabolism and performance. *Scandinavian Journal of Work, Environment & Health, 36*(2), 150–162.

Mirnalini, K., Zalilah, M.S., Safiah, M.Y., Tahir, A., Siti, H.M.D., Siti, R.D. & Normah, H. (2008). Energy and nutrient intakes: Findings from the Malaysian Adult Nutrition Survey (MANS). *Malaysian Journal of Nutrition, 14*(1), 1–24.

National Coordinating Committee on Food and Nutrition (NCCFN). (2005). *Recommended nutrient intakes for Malaysia.* Kuala Lumpur, Malaysia: Ministry of Health.

Sahu, S. & Dey, M. (2011). Changes in food intake pattern of nurses working in rapidly rotating shift. *Al Ameen Journal of Medical Sciences, 4*(1), 14–22.

Schaller, C. & James, E.L. (2005). The nutritional knowledge of Australian nurses. *Nurse Education Today, 25*, 405–412.

Strzemecka, J., Bojar, I., Strzemecka, E. & Owoc, A. (2014). Dietary habits among persons hired on shift work. *Annals of Agricultural and Environmental Medicine, 21*(1), 128–131.

Varli, S.N & Bilici, S. (2016). The nutritional status of nurses working shifts: A pilot study in Turkey. *Revista de Nutrição, 29*(4), 589–596.

Wong, H., Wong, M.C.S., Wong, S.Y.S. & Lee, A. (2010). The association between shift duty and abnormal eating behavior among nurses working in a major hospital: A cross-sectional study. *International Journal of Nursing Studies, 47*(8), 1021–1027.

*Strengthening Research Capacity and Disseminating New Findings in Nursing and Public Health – Malini et al. (Eds)*
*© 2018 Taylor & Francis Group, London, ISBN 978-1-138-50066-2*

# The effects of structured education on knowledge, attitude and action of patients with a colostomy

R. Fatmadona, H. Malini & M. Sudiarsih
*Faculty of Nursing, University of Andalas, West Sumatra, Padang, Indonesia*

ABSTRACT: The formation of colostomy causes problems; Most of the skin irritation around the colostomy is due to skin contact with impurities. Education on colostomy treatment can prevent this problem. The purpose of this study was to investigate the structured effect of education on colostomy care on knowledge, attitudes, and practices among patients at Padang General Hospital, Indonesia. Data collection was conducted in 2016, using an experimental quasi-study design, with the pretest-posttest single group. A total of 10 ostomates were chosen by purposive sampling. Statistics were performed with paired t-test with $p < 0.05$. The result confirmed the structured education had an effect on the knowledge, attitude, and practice of colostomy care. Sharing of knowledge and skills in colostomy care should exist as a routine agenda in preoperative nursing interventions and be provided in structured education programs for patients undergoing colostomy.

## 1 INTRODUCTION

A colostomy is a surgical procedure to create a hole through the abdominal wall into the iliac colon (ascending) as a stool outlet. It can be performed permanently or temporarily depending on the purpose of the operation (Nainggolan & Asrizal, 2013). The most common need for a colostomy is colorectal cancer. which is a malignant disease that attacks the colon (Ignatavicius & Workman, 2010).

Colorectal cancer is the third largest disease in the world (Siegel, et al, 2015), and its incidence is reported highly in developed countries, with 100,000 new ostomates added each year in the US (Davis, & Claudine, 2015). According to Dharmais Hospital in Indonesia, as a National cancer hospital, this cancer is ranked at the third top case in this hospital with a reported 269 new cases. This number is increasing with Indonesian people adopting unhealthy lifestyles. The growing number of colorectal cancer patients will, in turn, increase the number of colostomy sufferers (Data & Information Center, 2015).

The formation of a colostomy will cause many problems to the sufferers, physically, mentally, emotionally, socially, and economically (Panusur, & Nurhidayah, 2007). Some researchers found problems such as: skin irritation mainly from the leakage of urine or faeces; infection; and pyoderma gangrenosu; use of colostomy tools and accessories,; diet,; stoma, issues; psychological issues; and how to resume a normal life. It was found that these problems affected the patient's life in at least the first five years after having a colostomy fitted (Burch, 2013; Herlufsen et al., 2006; Jordan & Cristian, 2013). Those problems existed due to lack of knowledge and not seeking the help of health professional.

Colostomy patients should be taught how to manage their colostomy after the preoperative period so that the patient is able to perform a colostomy treatment independently once they are out of hospital (Burch, 2013). Patient education is a part of nursing care and provides an integrated health education centered on a patient's problems (Potter, & Perry, 2013). Educating patients can be undertaken in a structured and informal way. Nevertheless, structured education, according to the National Institute for Health and

Care Excellence (NICE, 2003), should be a planned and comprehensively rated education programme, flexible in content, responsive to an individual's clinical and psychological needs, and tailored to the education and cultural background of the patient. According to Kadam and Shinde (2014), the provision of structured education improves the knowledge and attitude of caregivers in performing colostomy care, while Danielson and Rosenberg (2014) stated, it could improve quality of life. The provision of education to patients will increase their responsibility for self-care and be able to carry out ongoing home care independently (Potter, & Perry, 2013). From the preliminary survey, four patients received only verbal, brief, and unstructured information about colostomy care. Stoma pouch and skin care were taught only when nurses performed surgical wound care and replaced a patient's' colostomy bag. Patients did not know when and how to change pockets, empty the right colostomy bag, care for the skin around the stoma to prevent irritation, what to do in case of irritation, and that certain foods could affect the elimination. Three people felt embarrassed and uncomfortable with having a colostomy fitted, the odour, and the discomfort felt. Based on these problems, further exploration and investigation on the effect of structured education on the knowledge, attitude, and actions of patients in the care of colostomies was deemed necessary.

## 2 METHOD

This was a quasi-experimental research involving one group with pre-test post-test design, giving treatment to the subjects, measuring and then analysing results of the treatment. The population were 72 ostomates being treated in Padang General Hospital during a least the last three months of 2016. Ten population samples were selected using a purposive sampling technique for the following criteria: temporary/permanent colostomy patients,; new ostomates,; post-op colostomy (four days or more),; and cooperative awareness.

The instruments consisted of three questionnaires and observation sheets. Questionnaire A elicitsed the characteristics of the respondents. Questionnaire B was used to assess patient knowledge with the Guttman scale., while questionnaire C was used to assess the patient's attitude with the Likert Scale. The results of the univariate data are displayed as a frequency distribution and percentage table. Bivariate analysis was used to test the relationship between the two variables studied. and data was processed by a computer program variable. The first data normality test was performed with a Shapiro-Wilk test, and then a t-paired test was used to test the hypothesis.

## 3 RESULTS

Prior to the intervention, a pre-test of the patients' knowledge and attitude on colostomy care using questionnaires B and C, was undertaken. Educational interventions were held for 30 minutes each session in the patient's room by using a booklet, counselling, and demonstration. After the intervention, a post-test on the same aspects was completed and evaluated.

In Table 1, most of respondents (70%) were $\geq$ 40 years old, 4 persons were elementary and middle school educated; 40% were housewives., 90% had a permanent colostomy fitted, and 80% had never received education on their stoma. More than half of the respondents (60%) had skin irritation around the colostomy site.

Table 2 shows that prior to the intervention, most respondents (60%) had low knowledge level and the intervention increased their knowledge so that 80% of them knew more. It was also found that 60% had negative attitudes before educational intervention, while their attitudes became 70% positive, after being given education (post-intervention). Lastly, more than half of respondents (60%) had been able to take good care of their colostomy treatment after being educated.

Table 1. Frequency distribution results based on respondent's' characteristics (n = 10).

| No. | Characteristics | Criteria | f | % |
|---|---|---|---|---|
| 1 | Age (years) | ≥40 | 7 | 70 |
| | | <40 | 3 | 30 |
| 2 | Gender | Male | 4 | 40 |
| | | Female | 6 | 60 |
| 3 | Education | Elementary | 4 | 40 |
| | | Middle school | 4 | 40 |
| | | High school | 1 | 10 |
| | | University | 1 | 10 |
| 4 | Working status | Farmer | 3 | 30 |
| | | NGO | 3 | 30 |
| | | Housewife | 4 | 40 |
| 5 | Economic level | Low | 1 | 10 |
| | | Moderate-low | 7 | 70 |
| | | Moderate | 11 | 10 |
| | | High | | 10 |
| 6 | Education on colostomy | Good | 2 | 20 |
| | | Bad | 8 | 80 |
| 7 | Colostomy type | Permanent | 9 | 90 |
| | | Temporary | 1 | 10 |
| 8 | Skin irritation | Yes | 6 | 60 |
| | | No | 4 | 40 |

Note: NGO = Non Government Officer.

Table 2. Frequency distribution based on respondents' knowledge, attitude and action on colostomy care (before and after), (n = 10).

| | Before | | After | |
|---|---|---|---|---|
| Item | f | % | f | % |
| Knowledge | | | | |
| High | 2 | 20 | 8 | 80 |
| Moderate | 2 | 20 | 2 | 20 |
| Low | 6 | 60 | 0 | 0 |
| Attitude | | | | |
| Positive | 4 | 40 | 7 | 70 |
| Negative | 6 | 60 | 3 | 30 |
| Action | – | – | | |
| Good | | | 6 | 60 |
| Not | | | 4 | 40 |

Table 3. Effect of knowledge of respondents before and after structured education about colostomy treatment.

| Knowledge | n | Mean | Increase | SD | *p* value |
|---|---|---|---|---|---|
| Before | 10 | 64,00 | 23,00 | 14,9 | 0,000 |
| After | 10 | 87,00 | | 8,9 | |

Table 3 shows that $p = 0{,}000$ ($p < 0.05$), then there is the effect of structured education on the patient's knowledge in colostomy care. Table 4 shows that the value of $p = 0.001$ ($p < 0.05$) confirming the hypothesis having an effect on the structured education on the patient's attitude in colostomy care is real.

Table 4. Effect of respondents' attitudes before and after structured education about colostomy treatment.

| Attitude | n | Mean rank | Increase | *SD* | *p* value |
|---|---|---|---|---|---|
| Before | 10 | 20,52 | 4,90 | 2,22 | 0,001 |
| After | 10 | 5,4 | | 4,06 | |

## 4 DISCUSSION

The results showed that the low level of knowledge of respondents, due to lack of information about colostomy treatment, where 60% of respondents have a low level of formal education. Education affects the learning process. The education provided will increase the acceptance of ostomates' responsibilities in self-care as well as enable them to carry out ongoing home—care independently (Potter, & Perry, 2013). After a structured education of colostomy care was held, most patients (80%) had a good knowledge and only a small proportion (20%) had sufficient knowledge. The statistical hypothesis test results confirmed the existence of the structured educational effect on changes in patient knowledge.

Health education is one of the processes to improve one's knowledge by obtaining more information from other people, printed mass media, and electronic sources, such as newspapers, leaflets, magazines, television and the radio (Mahdali, et al., 2013). Alenezi and Mansour (2016) also stated that the provision of structured education will improve patient knowledge in stoma care and reduce the incidence of peristomal skin complications when compared to non-educated patients. An interactive process that encourages the learning process is important, and learning is an effort to add new knowledge, attitudes and skills through strengthening certain practices and experiences (Widiastuti, 2013). In this study, a structured approach was taken in the provision of individual counselling and was undertaken for 30 minutes in two meetings using a booklet,; this was found to be very helpful and effective in delivering teaching materials and enhancing individual knowledge.

In this study, each respondent was given a booklet that could be taken home, so they are expected to understand and learn more at home. A contact number that could be used by the patient if they had any questions, was also given. The structured education was given through counselling and demonstrating the steps in colostomy care. The correct procedure of replacing the colostomy bag properly was demonstrated and the patient was then required to re-demonstrate what they had learnt. Moreover, this study discovered that there was no significant difference in attitudinal values between patients with permanent and temporary colostomies. This was due to the lack of certainty that temporary colostomy patients experienced. Hong, et al. (2014) also stated that there was no significant average difference in attitudinal values between patients with permanent and temporary colostomies, but there was little apparent difference in their assessment of attitudes towards worsening body image and lower self-esteem.

Attitude is the judgement (possibly the opinion) of a person against the stimulus or object. Once someone knows the stimulus and the object, the next process is to assess any change in the judgement against the stimulus/object (Ignatavicius, & Workman, 2010). A patient's attitude in this study indicated his/her knowledge of colostomy and its treatment. If a person's knowledge is not good, the resulting attitudes tend to be negative. This is simply because knowledge is one factor in the formation of an attitude. After respondents were given structured education, there was a positive attitude improvement from only 4 people (40%) to 7 people (70%). Improved attitudes in this study occurred because patients have gained knowledge through the provision of structured education about colostomy care. Moreover, the paired t-test result of $p = 0,001$ ($p < 0,05$) confirmed the influence of structured education on a respondents' change of attitude. The results of this study were similar to those of Kadam & Shinde (2014)who found an increase from 30% with a positive attitude to 70% having a strong positive attitude, after being given structured education on colostomy care. It can be concluded from the results and discussion that postoperative education intervention conducted for two days for the duration of only 30 minutes in each meeting, had not been able to significantly improve respondents' skills/ability in colostomy care.

## 4 CONCLUSION

It can be concluded that: there was a significant increase in respondents' knowledge before and after given structured colostomy care education; there was a significant improvement in respondents' attitudes before and after being given structured colostomy care education; patients had the ability to perform colostomy care after structured education; and structured education affected the patients' knowledge, attitude, and actions in the care of colostomy.

Hospitals need to include patient education into routine activities in every inpatient unit and to create a counselling division in each unit of care. Nevertheless, further research is needed to explore further actions for patients and their families in treating cancer patients.

## REFERENCES


Alenezi, A.N. & Mansour E.A., (2016), Impact of stoma care education in minimizing the incidence of stoma skin complications, *Bahrain Medical Bulletin, 38(3).*151–153.

Burch, J., (2013), Resuming a normal life: Holistic care of the person with an ostomy, *British Journal of Community Nursing, 16(8).* 366–373. Retrieved from https://www.ncbi.nlm.nih.gov/pubmed/21841628.

Danielson, A.K. & Rosenberg. J, (2014), Health related quality of life may increase when patient with a stoma attend patient education–A case control study. *PLoS One, 9(3).* Retrieved from http://journals.plos.org/plosone/article?id=10.1371/journal.pone.0090354.

Data & Information Center., (2015), *The incidence of colostomy,* Jakarta, Indonesia: Ministry of Health.

Davis, F. & Claudine, C., (2015), Physical disabilities: The role of occupational therapy in ostomy management for clients with cancer-related impairments. *Quality of Life Research Journal*, 25:, 125–133.

Hegazy, S.M., Ali, Z.H., Mahmoud, A.S., & Abou-zheid, A.A., (2014), Outcomes of educational guidelines on awareness and self efficacy among patients with permanent colostomy, *New York Science Journal, 7*(3). 25–32.

Herlufsen, P., Olsen A.G., Carlsen B., Nybaek, H., Karlsmark, T., Laursen, T. N., & Jemec, G. B., (2006), Study of peristomal skin disorders in patients with permanent stomas., *Pubmed, 15*(16), 854–862.

Hong, K.S., Oh, B.Y., Kim, E.J., Chung, S.S., Kim, K.H., & Lee, R.A., (2014), Psychological attitude to self-appraisal of stoma patients: Prospective observation of stoma duration effect to self-appraisal, *Annals of Surgical Treatment and Research, 86(3),* 152–156.

Ignatavicius, D.D., & Workman, M., L, (2010), *Medical-surgical nursing: Patient-centered colaborative care*, (6th ed.), Pennsylvania, Philadelphia: Elsevier/Saunders.

Jordan, R., & Cristian, M., (2013), *Understanding peristomal skin complications*. Retrieved from http://woundcareadvisor.com/understanding-periostomal-skin-complications_vol2_no3/.

Kadam, A., & Shinde, M.B., (2014), Effectiveness of structured education on care givers knowledge and attitude regarding colostomy care. *International Journal of Science and Research,3(4)* 586–593

Nainggolan, S.E., & Asrizal., (2013), *Educate family ability in stoma care at public hospital* (Master's thesis. Fkep USU, Medan, Indonesia).

National Institute for Health and Care Clinical Excellence (NICE), (2003), *Guidance on the use of patient-education models for diabetes technology appraisal.* Technology appraisal guidance [TA60] London, England: NICE Technology Appraisal Guidance.

Panusur, S. & Nurhidayah, M. E., (2007), *Self-care ability and self-image of colostomy patient at public hospital* (Master's thesis). Fkep USU, Medan, Indonesia.

Potter, P.A. & Perry, A. G., (2013), *Fundamentals of nursing*. Philadelphia, Pennsylvania: Elsevier. Inc.

Siegel, R.L., Miller, K.D., & Jemal, A., (2015), *Cancer Statistic, Jan–Feb; 65*(1):5–29.

Widiastuti, A. (2013). *The effectiveness of structured education based on the theory of planned behavior towards empowerment and quality of life of coronary heart disease patients in Pondok Indah hospital Jakarta.* (Master's thesis). University of Indonesia, Jakarta, Indonesia.

*Strengthening Research Capacity and Disseminating New Findings in Nursing and Public Health – Malini et al. (Eds)*
 *ISBN 978-1-138-50066-2*

# Nurse experiences in conducting diabetic foot risk assessment in diabetic patients in hospital in Medan

Y. Ariani & R. Tarigan
*Faculty of Nursing, University of Sumatra Utara, Medan, Indonesia*

ABSTRACT: Nurses have an important role in preventing the occurrence of diabetic foot injuries by conducting diabetic foot screening. Diabetic foot screening is the part of the nursing process that assesses the risk of diabetic foot. This study aims to explore the experience of nurses in assessing diabetic foot risk in diabetic patients in H. Adam Malik Hospital Medan. This research used qualitative design with a phenomenology approach with an in-depth interview of ten nurses working with inpatients. The interviews were transcribed precisely and then analysed to develop the research themes. The results of the study found three themes of nurse's experiences: (1) wound assessment; (2) lack of knowledge about diabetic foot screening; (3) diabetic foot training. Therefore, it is necessary to increase the knowledge and skill of nurses in conducting diabetic foot screening through training, workshop and simulation directly on patients.

## 1 INTRODUCTION

Diabetes is a chronic metabolic disease characterised by high blood glucose levels as a result of abnormalities in insulin secretion, impaired insulin activity, or both, that can cause serious problems and increase their levels rapidly (LeMone & Burke, 2008; Smeltzer & Bare, 2010). Data from the International Diabetes Federation (IDF) states that the amount of diabetic patients worldwide up to 2013 reached 382 million people and is predicted to increase by 55% by 2035, an estimated 592 million people. Epidemiologically diabetes is often undetectable and the onset of the disease is seven years before diagnosis is established (IDF, 2013). Indonesia is the seventh out of ten countries with the highest number of diabetes cases at the age of 20–79 years and the number of patients reaching 8.5 million people (IDF, 2013).

Based on data from National Diabetes Fact Sheet (2011) about 60–70% of diabetics suffer complications of mild to severe neuropathy that will result in loss of sensory and damage to leg. Forlee (2010) stated that 15% of patients will have diabetic foot ulcers and up to 70% will have amputations. The mortality and amputation rates of diabetic foot ulcers are also high at 16% for mortality and 25% for amputation rates (Sudoyo et al., 2009). Diabetes is thought to be the main causative factor in 45% of all lower extremity amputations, with 60% of non-traumatic amputations (Heitzman, 2010).

Given the increasing number of diabetic patients and the high cost of patient care that is mainly caused by complications. One of the prevention efforts is done by assessing of diabetic foot. The assessing and screening of the diabetic foot should be performed by the nurse at the time the patient is first treated and during treatment of the disease. The purpose of this study is to explore the experience of nurses in conducting diabetic foot screening in Diabetic patients in H. Adam Malik Hospital Medan.

## 2 METHODS

This research used qualitative design with phenomenology approach, with the aim to obtain experience from nurses about diabetic foot assessment in hospital. This research used qualitative design with phenomenology approach with an in-depth interview with nurses working with inpatients. The researchers conducted an interview (in-depth interview) for 30–60 minutes with the nurse in the hospital room in H. Adam Malik Hospital Medan until sufficient data was obtained from ten nurses.

This study was conducted using ethical guidelines of self-determination, privacy, anonymity, informed consent and protection from discomfort research.

## 3 RESULTS

Table 1 shows that half (50%) of the nurses are 36–45 years old, the majority (90%) are female, 60% are nursing undergraduate, 60% have worked for 11–20 years and only 30% have ever received training on diabetes.

The results of the study found three themes of nurse experience in assessing diabetic foot: (1) wound assessment; (2) lack of knowledge about diabetic foot screening; and (3) diabetic foot training.

Themes 1. Wound assessment

The majority of nurses say that the most important diabetic foot assessment is examining the wound, as shown from the following statement of some participants:

> All this time, we assess the patient's feet while taking care of the wound, we measured the extent of the wound, assess the colour of the wound, there are still many exudates and new tissue growing
>
> (Participant 1)
>
> We need to examine the wounds, because the majority of them come here already with bad wounds
>
> (Participant 3)
>
> Assessment of the foot ... it examines the wound, assess the development of the wound, check blood glucose level
>
> (Participant 7)

Table 1. Distribution of frequency and percentage of nurse demographic data at H. Adam Malik Hospital Medan (n = 10).

| Characteristics of respondent | Frequency (f) | Percentage (%) |
|---|---|---|
| Ages | | |
| 25–35 years | 3 | 30 |
| 36–45 years | 5 | 50 |
| >46 years | 2 | 20 |
| Gender | | |
| Female | 9 | 90 |
| Male | 1 | 10 |
| Education | | |
| Nursing diploma | 4 | 40 |
| Nursing undergraduate | 6 | 60 |
| Length of work | | |
| 1–10years | 3 | 30 |
| 11–20 years | 6 | 60 |
| >20 years | 1 | 10 |
| Diabetic training | | |
| Ever | 3 | 30 |
| Never | 7 | 70 |

Themes 2. Lack of knowledge about diabetic foot screening

In addition to the absence of a diabetic foot screening format, nurses are also less knowledgeable about it. Nurses know less about assessing foot deformities, sensory assessment, vascular assessment and ABI (Ankle Brachial Index). This can be seen from some participant statements:

> We recommend that both legs examined, assess signs of infection, fungus in between toes, history of ulcers.
>
> (Participant 4)
>
> Because it was never done less know what to examine for this foot screening ... most touched the skin of his legs, asked what still feels when held on his toes, seen any signs of inflammation or risk of injury
>
> (Participant 6)
>
> Never touched the dorsal pedis artery ... never compared also left and right
>
> (Participant 7)
>
> What ABI? ... ooo ... we never do it, we don't know how, no tool in room, I've seen residents doctors take the tool but did not have time to pay attention to that time for caring for other patients
>
> (Participant 10)
>
> Which kind of measure blood pressure in the foot, if in the arm ... we can do it ... wear a special tool, like measure ordinary blood pressure?
>
> (Participant 9)

Theme 3. Diabetic foot training

The majority of nurses have never received training on diabetic foot screening. Only two have ever received diabetes training, although only general, and one person has received wound care training. Here are the nurse statements:

> I have never participated in diabetic foot screening training, if general diabetic training has been done, but not specifically about diabetic foot screening.
>
> (Participant 1)
>
> If the wound training I just joined one month ago in this hospital, but reviewing the foot as the you said earlier, assess sensory, measure the value of ABI, distinguish the shape of the foot ... there is don't ...
>
> (Participant 3)
>
> I have never participated in any diabetes-related training, and I've been working here for three years now ... can be proposed for diabetic foot screening training
>
> (Participant 8)
>
> I think, we don't know about it caused we not being exposed to diabetic foot screening format, work rotation and non-nursing tasks
>
> (Participant 10)

## 4 DISCUSSION

Nurses have an important role in supporting and maintaining the healthy feet of a diabetic patient (Delmas, 2006 in Ariyanti, 2012). Nurses are required to conduct a comprehensive review, plan, nursing interventions, collaborative interventions, evaluation and follow-up. The role of nurses in providing nursing care to DM patients, especially in preventing diabetic foot ulcer complications, by examining diabetic foot, determining the risk of diabetic foot (diabetic foot screening) and follow-up for treatment (Boulton et al., 2008).

Early identification of patients at high risk of diabetic foot ulcers is a high priority in diabetics care as it can reduce the complications and economic burden of diabetic foot complications. Regular screening of patients at high risk of diabetic foot is an important step as a reference of care, prevention and optimisation of health resource utilisation. 85% of the amputations associated with lower limb diabetes are preceded by diabetic foot ulcers, therefore routine screening is essential (Reiber, Lipsky & Gibbons, 1998). A study by Narayan et al., (2006) showed that prevention by diabetic foot screening in high-risk patients was one of the three best ways of saving costs and diabetic intervention, while the others were glycemic control to achieve HbA1c < 9.0% and blood pressure control to achieve < 160/95 mmHg.

The purpose of diabetic foot screening is to identify foot problems, determine foot risk categories, patient management categories and to instruct patients and families in appropriate foot care (Yetzer, 2004). The nurse in H. Adam Malik Hospital stated that they had not understood and have little idea how to conduct diabetic foot screening or the purpose of diabetic foot screening. Nurses are more focused on assessing and treating diabetic foot wounds. The absence of a diabetic foot screening format in hospitals also causes nurses not to screen diabetic foot.

The diabetic foot assessment format has not been applied in Adam Malik Hospital Medan. The format of diabetic foot screening should be standardised. A standardised format is the complete documentation of legal nursing care in the application of nursing practice at the clinic. This format shows the nurse's performance that accurately assesses the progress of the treated patient; the data in this format is a benchmark for evaluating whether the patient has improved or deteriorated as well as the benchmark for successful nurse performance in treating the patient (Woodbury et al., 2015)

Lack of nurses' knowledge of diabetic foot screening is related to all of the participating nurses in this study had never received training on diabetic foot screening, they said that they still at least training on diabetics care. The length of work experience does not affect the knowledge and skills of nurses, whereas 60% of nurses in this study have been working for more than ten years.

The nurses also do not know about the tools used for screening sensory function changes in the legs. The average nurse does not understand the assessment of the location of arterial pulsation in the legs. This condition may be related to the need for continuous improvement in science and skills. The development of science and technology develops periodically and continuously, so the nurse must be able to follow these developments by following the latest developments in science and skills. The knowledge and skills that the nurse must know in assessing the diabetic foot are the patient history, general inspection, dermatology assessment, musculoskeletal assessment, neurological assessment and vascularisation assessment (Boulton et al., 2008).

Historical assessment is very important in determining risk factors and should be followed by foot examination. The general inspection of diabetic foot such as foot size is very important, because shoes and socks that do not fit the foot size are often the cause of diabetic foot ulcers. Therefore it is necessary to check the suitability and comfort of shoes and socks with foot sizes. Skin assessment is carried out by examining the skin of the foot, fold the toes to see ulceration or erythema. The musculoskeletal study aims to see if there is foot deformity. Rigid deformity is usually found in the toes. Foot deformity can increase the pressure on the plantar and result in damage to skin integrity, including metatarsal phalangeal hypersensitivity with interphalangeal flexion or distal extension phalangeal (hammer toe) (Miller, et al., 2014).

Peripheral neuropathy is the most common cause factor in the occurrence of diabetic foot ulcers. Clinical examination recommends identifying the Loss of Protective Sensation (LOPS) or loss of sensation protection, using 10 g monofilament to assess the sensation. In addition, it is important to conduct assessment of peripheral artery disease in determining the risk status of DM legs. Vascular examination by palpation of the tibial posterior pulse and, with the results "palpable" or "not palpable" pulse. Patients with signs and symptoms of vascular disease or pulsation disorders in foot screening should be examined using the Ankle Brachial Index (ABI). ABI examination compares the systole blood pressure value in ankle with in brachialis. If the ABI value > 0.9 it is normal, < 0.8 are associated claudication and < 0.4 are associated with ischemic rest pain and tissue necrosis (Singh et al., 2005). From the diabetic foot screening results and patient compliance in implementing the follow-up recommendations, the expected incidence of diabetic foot injuries and the risk of amputation in DM patients can be prevented as early as possible.

## 5 CONCLUSIONS

The study concluded that nurses focus solely on assessing diabetic foot injuries, not assessing the risk of diabetic foot, and have a lack of knowledge about diabetic foot screening and the need for diabetic foot screening training. It is recommended that the knowledge and skills of nurses in carrying out diabetic foot screening through training is increased, workshop and direct simulation on patient in diabetic foot examination so that the patient's assessment becomes more perfect, complete the existing wound assessment and nursing care in DM patient become more comprehensive and precise.

## REFERENCES


Ariyanti, (2012). Relationship of foot care with risk diabetic foot ulcers in PKU Muhammadiyah Hospital Yogyakarta. ***Tesis FIK U,*** *page 1–95.* Retrieved from http://lib.ui.ac.id/file?file=digital/20308399-T31066-Hubungan%20perawatan.pdf.

Boulton, A.J.M., Amstrong, D.G., Albert, S.F., Fryberg, R.G., Hellman, R. & Kirkman, M.S. (2008). Comprehensive foot examination and risk assessment. *DM Care Journal. 31*(8). 1679–1685.

Forlee, M. (2010).What is the diabetic foot? *CME, 28*(4), 152–156.

Heitzman, J. (2010). Foot care for patients with DM. *Topics in Geriatric Rehabilitation, 26*(3), 188–191.

IDF. (2013). *IDF DM Atlas* (6th ed.). The International DM Federation. Retrieved from *https://www.idf.org.*

LeMone, P.T. & Burke, K.M. (2008). *Medical surgical nursing: Critical thinking in client care* (4th ed.). New Jersey: Pearson Prentice Hall.

Miller, J.D., Carter, E. & Shih, J. (2014). How to do a 3-minute diabetic foot exam. *The Journal of Family Practice, 63*(11), 646–649, 653–656.

National DM Fact Sheet. (2011). *Fast Fact on Diabetes.* Retrieved from http://www.cdc.gov/diabetes/pub/pdf/ndfs-2011.

Narayan, K.V., Zhang, P., Kanaya, A.M., Williams, D.E., Engelgau, M.M. & Imperatore, G. (2006). *Diabetes: The pandemic and potential solutions. Disease control priorities in developing countries* (2nd ed.). Washington D.C.: World Bank. Retrieved from http://www.ncbi.nlm.nih.gov/books/NBK11777/.

Reiber, G., Lipsky, B., & Gibbons, G. (1998). The burden of diabetic foot ulcers. *Am J Surg, 176*(2), 5S–10S.

Smeltzer, S.C., & Bare, B.G. (2010). *Brunner & Suddarth's textbook of medical surgical nursing.* Philadelphia: Lippincott.

Singh, N., Amstrong, D.G. & Lipsky, B.A. (2005). Preventing foot ulcers in patients with DM. *Journal of the American Medical Association, 293*(2), 217–228. doi:10.1001/jama.293.2.217.

Sudoyo, A.W., Setiyohadi, B., Alwi, I., Simadibrata, M. & Setiati, S. (2009). *Internal medicine textbook.* Ed5. Jakarta: Interna Publishing.

Woodbury, M.G., Sibbald, R.G., Ostrow, B., Persaud, R. & Lowe J.M. (2015). Tool for rapid & easy identification of high risk diabetic foot: Validation & clinical pilot of the simplified 60 second diabetic foot screening tool. *PLoS ONE 10*(6), e0125578. doi:10.1371/journal.pone.0125578.

Yetzer, E.A. (2004). Incorporating foot care education into diabetic foot screening. *Rehabilitation Nursing 29*(3), 80–84.

*Nursing education*

*Strengthening Research Capacity and Disseminating New Findings in Nursing and Public Health – Malini et al. (Eds)*
 *ISBN 978-1-138-50066-2*

# The relationship between admission factors and first-semester grade point average in Indonesian nursing students

C.L. Sommers
*Universitas Pelita Harapan, Tangerang, Indonesia*

S.H. Park
*University of Kansas, Kansas City, Kansas, USA*

ABSTRACT: The purpose of this study was to determine what experience factors (region of origin and attendance at a pre-nursing course), attribute factors (gender) and academic metric factors (admission exam scores and type of high school) are associated with the first-semester Grade Point Average (GPA) of first-year nursing students enrolled in a baccalaureate nursing programme in Indonesia, using an adapted holistic admission model. A descriptive correlation design was used. Multivariate linear regression was used to determine the relationship between factors and first-semester GPA. A significant relationship ($p < 0.05$) was found between experience factors, attribute factors and academic metric factors and first-semester GPA. However, the study variables only accounted for 28% of the variance in GPA. Additional research is needed to identify other factors, guided by the adapted holistic admission model, that may also have a relationship with GPA.

## 1 INTRODUCTION

As Indonesia is the fourth most populous country in the world with 252.8 million people in 2014 (World Bank Group, 2015) and has a high demand for health care services, primarily supported by nurses, it is important that nursing programmes are preparing graduates to meet the demand in terms of quantity and quality of graduates. Holistic admissions review is a growing movement in healthcare education, including nursing, as an individualised way of assessing potential student's capabilities to contribute value to a healthcare profession that includes a balanced consideration of various factors (American Association of Medical Colleges, 2013; Scott & Zerwic, 2015). Scott and Zerwic (2105) described the adaptation of the medical model of holistic admission for use in nursing to increase the diversity among students and included factors of experiences (i.e. experiences in life, education, leadership and cultural diversity), attributes (i.e. demographics, maturity and goals) and academic metrics (i.e. grades and pre-admission to healthcare programme test scores). According to the model, balanced consideration is given to these factors, so that the emphasis is on the contribution the applicant will make to the nursing profession (Scott & Zerwic, 2015).

As holistic admissions review is implemented in nursing programmes, it will be important that academic success is monitored (Glazer et al., 2016). A strong predictor of graduation is a student's academic achievement, which includes grades (American Council on Education, 2016). However, in nursing, there is a lack of data on which student factors best predict future success (Glazer et al., 2016). It is important to determine what factors may be related to academic achievement of nursing students. The adapted holistic admission model (Scott & Zerwic, 2015) is useful for guiding the selection of factors that may affect Grade Point Average (GPA).

The purpose of this study was to determine what experience factors (region of origin and attendance at a pre-nursing course), attribute factors (gender) and academic metric factors (admission exam scores and type of high school) are associated with the first-semester GPA of nursing students enrolled in a baccalaureate programme in Indonesia. First-semester GPA has been associated with success in nursing programmes (Newton & Moore, 2009).

## 2 METHOD

### 2.1 *Study design and sample*

A descriptive correlation design using secondary data analysis was used. The data set was from academic and admission data that were collected in a database of all accepted first-year nursing students at an Indonesian university between August and December 2016. A power analysis was conducted to determine the minimum sample size needed (Green, 1991). With 12 variables, a medium effect size, alpha level set at 0.05 and power set at 0.80, the minimum sample size was 127. The size of the data set was 510, which exceeded minimum sample size.

### 2.2 *Ethical consideration*

The privacy of students was protected using a de-identified database that did not contain any student identifying information. All data in the de-identified database was kept confidential and stored on secure servers accessed from password-protected computers. Approval for the study was from the Mochtar Riady Institute of Nanotechnology Ethics Committee in Indonesia and from the University of Kansas Medical Centre Human Research Protection Program in the US.

### 2.3 *Data collection and analysis*

The variables of interest for this study were first-semester GPA, the experience factors of region of origin and attendance at a pre-nursing course, the attribute factor of gender, and the academic metric factors of admission exam scores and type of high school. A previous study has shown a relationship between GPA in the first semester and completion of the nursing study programme (Newton & Moore, 2009). Bacon and Bean (2006) determined that the reliability of first-year GPA with all courses was 0.84 and by the end of the fourth year, it had increased to 0.94. They suggest that GPA is a reliable indicator to measure academic performance of students.

The variables of region of origin and attendance at a pre-nursing course were chosen as experience factors for this study. Region of origin was used to represent various experiences in culture and life and was defined as the region in Indonesia where the student resided before coming to university and was divided into five categories (See Table 1). The variable attendance at a pre-nursing course was defined as attendance, by invitation, to come before the start of the first semester for intense instruction in life skills, basic computer and math skills, and introduction to English. A previous study found a weak association with region of origin/ethnicity and GPA in nursing students in New Zealand (Shulruf et al., 2011). However, little is known about the relationship between experiences and GPA for Indonesian nursing students.

The variable of gender, defined by the World Health Organization in 2011 as socially constructed characteristics of women and men, is an attribute factor. Previous studies have not found a relationship between gender and GPA (Shulruf et al., 2011). It is unknown what the relationship may be in nursing students in Indonesia.

Academic metrics included type of high school and scores on four admission exams. The type of high school was defined as the classification of the high school programme the student attended. The types of high schools were divided into three categories (See Table 1). The exams for math, English and Indonesian were developed by faculty members that taught those subjects. No reliability and validity studies have been conducted. The math exam had a range of 0–30 and tested basic math concepts. The English exam had a range of 0–50 and tested English reading ability and grammar. The Indonesian exam had a range of 0–40 and tested Indonesian reading ability and grammar. The logic patterns exam is the language-free version of Raven's

Table 1. Characteristics of the sample ($N$ = 506).

| Variable | Mean | Standard deviation |
|---|---|---|
| Admission exam scores (possible range) | | |
| Math admission exam score (0–30) | 10.17 | 3.91 |
| English admission exam score (0–50) | 19.93 | 5.36 |
| Indonesian admission exam score (0–40) | 17.23 | 3.60 |
| APM admission exam score (0–36) | 21.55 | 3.55 |
| GPA (0.00–4.00) | 3.03 | 0.25 |
| | *n* | % |
| Region of origin | | |
| Sumatra Island (reference group) | 161 | 31.8 |
| Java and Bali Islands | 106 | 20.9 |
| Eastern Islands | 104 | 20.6 |
| Sulawesi Island | 94 | 18.6 |
| Kalimantan Island | 41 | 8.1 |
| Attendance at pre-nursing course | | |
| No (reference group) | 394 | 77.9 |
| Yes | 112 | 22.1 |
| Gender | | |
| Female (reference group) | 401 | 79.2 |
| Male | 105 | 20.8 |
| Type of high school | | |
| Science high school (reference group) | 402 | 79.4 |
| Social science vocational or high school | 60 | 11.9 |
| Health/science vocational school | 44 | 8.7 |

Advanced Progressive Matrices (APM) licensed by the University of Indonesia. The international technical manual (Raven, 2011) states that the reliability of the APM for studies done in the United States, as measured by split-half internal consistency, was 0.85, indicating good reliability because it is greater than 0.80 (Polit & Beck, 2017). The manual also discusses evidence of content, convergent and criterion validity and recommends that local validity studies be completed. Nursing programme admission exam scores have been associated with the GPA of first-year nursing students (Shulruf et al., 2011; Underwood et al., 2013).

Data were analysed using SPSS version 21statistical software. Descriptive statistical analysis included means and standard deviations for continuous variables, and frequency distributions and percentages for categorical variables. Multivariate linear regression was used to determine the relationship between the variables and first-semester GPA. The data were first explored for missing data and whether they met the assumptions of linearity, normality, non-multicollinearity, hemodesckeditiy and independence. Four of the participants had missing data related to type of high school, a missing data rate of 0.8%. Since the missing data was less than 1%, listwise deletion was used and it is acknowledged that there is a small potential for bias (Parent, 2013). After listwise deletion, the remaining data ($N$ = 506) were examined for meeting the assumptions of multivariate linear regression and all assumptions were met.

## 3 RESULTS

The characteristics of the final sample of 506 participants are displayed in Table 1. The mean GPA of the sample was 3.03. Most of the participants were from the island of Sumatra and did not attend the pre-nursing course. There were 401 females (79.2%) and 105 males (20.8%). The means of the admission exams were: math 10.17; English 19.93; Indonesian 17.23; APM 21.55. Most of the participants attended a science-focused high school.

Table 2. Regression table.

| Variable | B | SE(B) | 95% CI for B Lower | Upper | *t* | Sig. | β |
|---|---|---|---|---|---|---|---|
| Gender | −0.07 | 0.02 | −0.12 | −0.03 | −3.13 | 0.002 | −0.12 |
| Attendance at pre-nursing course | −0.01 | 0.03 | −0.05 | 0.04 | −0.22 | 0.827 | −0.01 |
| Java and Bali Islands (dummy variable) | −0.06 | 0.03 | −0.11 | −0.00 | −2.08 | 0.038 | −0.10 |
| Kalimantan Island (dummy variable) | −0.01 | 0.04 | −0.09 | 0.06 | −0.39 | 0.701 | −0.02 |
| Sulawesi Island (dummy variable) | −0.03 | 0.03 | −0.09 | 0.02 | −1.19 | 0.236 | −0.05 |
| Eastern Islands (dummy variable) | −0.03 | 0.03 | −0.08 | 0.03 | −0.95 | 0.345 | −0.04 |
| English admission exam score | 0.01 | 0.00 | 0.01 | 0.02 | 6.09 | 0.000 | 0.27 |
| Math admission exam score | 0.01 | 0.00 | 0.00 | 0.01 | 2.28 | 0.023 | 0.10 |
| Indonesian admission exam score | 0.01 | 0.00 | 0.01 | 0.02 | 5.04 | 0.000 | 0.21 |
| APM admission exam score | 0.01 | 0.00 | 0.00 | 0.01 | 3.27 | 0.001 | 0.13 |
| Health/science vocational school (dummy variable) | −0.02 | 0.04 | −0.10 | 0.05 | −0.67 | 0.503 | −0.03 |
| Social studies high school or vocational school (dummy variable) | −0.10 | 0.03 | −0.16 | −0.04 | −3.20 | 0.001 | −0.13 |

*Note:* SE = Standard Error; CI = Confidence Interval; Sig. = *t*-test significance.

Blockwise multivariate linear regression was done to determine the relationship between GPA and the independent variables. This study performed a three-block model (Model 1) and a two-block model (Model 2). Model 1 blocked the variables in three steps: (1) attribute factors; (2) experience factors; (3) academic metric factors. Model 2 had two steps: (1) attribute and experience factors; (2) academic metric factors. Only 2% of the variance was explained by attribute factors in Model 1. Thus, Model 2, including both attribute and experience factors in the first block, was chosen and reported in this study. The multiple correlation coefficient $R$ was 0.53 for the final model. The $R^2$ was 0.28, indicating that about 28% of the variance in GPA was accounted for by all independent variables, and only 5% of the variance was explained by attribute and experience factors. This linear combination of independent variables was significantly associated with GPA, $F(6, 493) = 26.62$, $p < 0.001$.

Based on analysis of beta coefficients, several independent variables were associated with GPA (see Table 2). When individual variables using standardised beta scores were examined, the score on English admission exam explained the most variance in GPA; followed by the score on Indonesian admission exam; score on APM admission exam; type of high school, social science focused vocational or high school when compared to science-focused high school; gender, male when compared to female; score on the math admission exam; and region of origin, islands of Java and Bali when compared to island of Sumatra. There was no significant relationship between attendance at the pre-nursing course and GPA.

Controlling for all the other variables, when each of the admission exam scores was increased by one point, the GPA increased by 0.01. If the student attended a social science focused vocational or high school, the GPA decreased by 0.10, compared to students that attended a science high school. Male students had a 0.07 lower GPA than female students. Students from the islands of Java and Bali had a 0.06 lower GPA than students from the island of Sumatra.

## 4 DISCUSSION

This study examined what experience factors (region of origin and attendance at pre-nursing course), attribute factors (gender) and academic metric factors (admission exam scores and type of high school) were associated with the first-semester GPA of first-year nursing students in a baccalaureate nursing programme in Indonesia. A significant relationship ($p < .05$) was found between region of origin, gender, admission exam scores, type of high school and first-semester

GPA. No significant relationship was found with attendance at the pre-nursing course. This may be because the pre-nursing course provided additional learning and support before the first semester and assisted in preparing those students for the rigours of the nursing programme.

The findings of a relationship between region of origin and GPA is similar to previous findings (Shulruf et al., 2011). It was surprising that students from the islands of Java and Bali had significantly lower GPAs, as those islands tend to have stronger high school programmes. Perhaps the region of origin needs to be divided into more categories to better explore association with GPA. Previous studies have also found a relationship between admission exam scores and GPA (Shulruf et al., 2011; Underwood et al., 2013). The findings in this study of a relationship between gender and GPA was not found in previous studies (Shulruf et al., 2011). Students that did not attend a science-focused high school had a lower GPA at the end of the first semester.

The admission exam scores for English and Indonesian had the most influence on first-semester GPA. This may be because students were enrolled in a General English course in the first semester and those that scored higher on the admission English exam may also have achieved a higher grade in the General English course, resulting in a higher GPA. It may also be that those that scored higher on the Indonesian admission exam have a better Indonesian reading and writing ability that influenced their general performance in all courses, resulting in a higher GPA.

## 5 CONCLUSIONS

This study adds to the body of knowledge related to the holistic admission model. Factors of experience, attributes and academic metrics were found to have an association with first-semester GPA. The findings that male students, students from a social science vocational or high school, and students from the islands of Java or Bali had lower GPAs have meaningful implications in identifying students at possible high-risk of a GPA below 2.75. As these students may be at risk of a lower GPA and being unsuccessful in the nursing programme, student support interventions, such as academic counselling, study skills workshops, writing resources, support groups and so on, could be applied to assist them during the first semester and throughout the nursing programme. Identifying high-risk students early and implementing support interventions may assist the students to improve their GPA and successfully complete the nursing programme.

As the variables in this study only explained 28% of the total variance in GPA, additional research is needed to determine what other variables are associated with first-semester GPA in first-year nursing students (i.e. interviews, psychology test results, support services, study habits, etc.). The variables may be a combination of admission factors (i.e. experience, attributes and academic metrics) and factors that occur during the first semester (i.e. tutoring, transition to university, study habits). Identifying such factors may aid in early identification of students that are at high-risk of not completing the nursing programme.

## REFERENCES

American Association of Medical Colleges. (2013). *AAMC Holistic Review Project*. Washington, DC: AAMC. Retrieved from https://www.aamc.org/initiatives/holisticreview/.

American Council on Education. (2016). *Unpacking relationships: Instruction and student outcomes*. Washington, DC: American Council on Education. Retrieved from http://www.acenet.edu/news-room/Documents/Unpacking-Relationships-Instruction-and-Student-Outcomes.pdf.

Bacon, D.R. & Bean, B. (2006). GPA in research studies: An invaluable but neglected opportunity. *Journal of Marketing Education*, *28*, 35–42. doi:10.1177/0273475305284638.

Glazer, G., Clark, A., Bankston, K., Danek, J., Fair, M. & Michaels, J. (2016). Holistic admissions in nursing: We can do this. *Journal of Professional Nursing*, *32*, 306–313.

Green, S.B. (1991). How many subjects does it take to do a regression analysis? *Multivariate Behavioral Research*, *26*(3), 499–510.

Newton, S.E. & Moore, G. (2009). Use of aptitude to understand bachelor of science in nursing student attrition and readiness for the National Council Licensure Examination-Registered Nurse. *Journal of Professional Nursing, 25*(5), 273–278.
Parent, M.C. (2013). Handling item-level missing data. *The Counseling Psychologist, 41*, 568–600.
Polit, D.F. & Beck, C.T. (2017). *Nursing research: Generating and assessing evidence for nursing practice.* Philadelphia, PA: Wolters Kluwer.
Raven, J.C. (2011). *Raven's Advanced Progressive Matrices: International technical manual.* San Antonio, TX: Pearson Education.
Scott, L.D. & Zerwic, J. (2015). Holistic review in admissions: A strategy to diversify the nursing workforce. *Nursing Outlook, 63*, 488–495.
Shulruf, B., Wang, Y.G., Zhao, Y.J. & Baker, H. (2011). Rethinking the admission criteria to nursing school. *Nurse Education Today, 31*(8), 727–732.
Underwood, L.M., Williams, L.L., Lee, M.B. & Brunnert, K.A. (2013). Predicting baccalaureate nursing students' first-semester outcomes: HESI admission assessment. *Journal of Professional Nursing, 29*(2 Suppl 1), S38–S42.
World Bank Group. (2015). *Total population (in number of people).* Washington, DC: The World Bank. Retrieved from http://data.worldbank.org/indicator/SP.POP.TOTL.
World Health Organization. (2011). *Gender, equity and human rights.* Geneva, Switzerland: World Health Organization. Retrieved from http://www.who.int/gender-equity-rights/knowledge/glossary/en/

*Strengthening Research Capacity and Disseminating New Findings in Nursing and Public Health – Malini et al. (Eds)*
*© 2018 Taylor & Francis Group, London, ISBN 978-1-138-50066-2*

# A systematic review: A tool development process for public health nurses' competencies

I. Kusumaningsih
*Sint Carolus School of Health Sciences, Jakarta, Indonesia*

A. Talosig
*St. Paul University Philippines, Cagayan, Philippines*

ABSTRACT: The availability of self-reporting tools with acceptable psychometric properties may contribute to helping researchers understand the tools' development. The purpose of this study was to systematically review the existing literature to identify the sustaining indices of a successful development and validation of a tool to support public health nurses' competencies. An integrative review of literature was conducted using keyword searches in ProQuest®. From 50 articles found, seven articles were reviewed that met both inclusion and exclusion criteria. Each tool specifies a unique set of dimensions of public health nursing competencies, the steps to develop the tool, and its validity and reliability results. Two themes that emerged were the process of developing a tool and the public health nursing competencies. The findings of this study will guide and facilitate deeper insights, contribute to the repository of knowledge, and propose a process for development of tools relating to public health nursing competencies.

*Keywords*: tool development, public health nurse, competency

## 1 INTRODUCTION

In order to maintain a community's health, public health nurses will make efforts in primary prevention, secondary prevention, and tertiary prevention (Lawson, 2014). Public health nursing uses subjective methods through professional judgements, and objective information through available data or specifically designed surveys in order to assess the needs of individuals, families, and the community (Sakellari, 2012; Rowley, 2005). Philibin et al. (2010) revealed that the role of Irish public health nurses emerged in four themes, which were: 'Jack of all trades: The role of the public health nurse defined and described', 'The essence of the role', 'Challenges to the role of the public health nurse', and 'Communication'. Kotrotsiou et al. (2008) emphasised that in the framework of nursing, nurses also play a counselling role. The role of the nurse with counselling abilities is to rehabilitate the patient physically, spiritually, and psychologically, and to assist them to regain their previous personal and social roles in the best possible way. A competent public health workforce effectively performs the ten essential public health services as outlined in Table 1 (CDC, 2014).

Prevost et al. (2015) contended that the health profession must demonstrate sufficient competencies in their health sector to become professionals. Thus, nurses, including public health nurses, need to be competent in implementing nursing care. Hurd and Buschbom (2010) defined competency as the skills, knowledge, and personal characteristics needed for

Table 1. Ten essential public health services (adapted from CDC, 2014).

| Core function | Essential service |
|---|---|
| Assessment | 1. Monitor health status to identify and solve community health problems |
| | 2. Diagnose and investigate health problems and health hazards in the community |
| Policy development | 3. Inform, educate, and empower people in relation to health issues |
| | 4. Mobilise community partnership and action to identify and solve health problems |
| | 5. Develop policies and plans that support individual and community health efforts |
| Assurance | 6. Enforce laws and regulations that protect health and ensure safety |
| | 7. Link people to needed personal health services and assure the provision of health care when otherwise unavailable |
| | 8. Assure competent public and personal health care of workforce |
| | 9. Evaluate effectiveness, accessibility, and quality of personal and population-based health services |
| | 10. Research new insights and innovative solutions to health problems |

successful performance in a job. However, Halcomb et al. (2016) argued that there are a limited number of published literature reports on competency standards for nurses working in general practice and primary health care.

In the US, the Quad Council alliance of four national organisations concerned with Public Health Nursing (PHN) took the Council on Linkages (COL) between Academia and Public Health Practice 'Core Competencies for Public Health Professionals' and applied it to two levels of PHN practice: the staff nurse/generalist role and the manager/specialist/consultant role. These Quad Council competencies were organised into eight domains, reflecting skill areas of public health and PHN practice: Analytics and Assessment; Policy Development/ Programme Planning; Communication; Cultural Competency; Community Dimensions of Practice; Public Health Service; Financial Management and Planning; Leadership and Systems Thinking (Swider et al., 2013).

## 2 PURPOSE

The purpose of this study was to identify and to integrate the findings of all relevant studies of tool development for PHN competency by reviewing the existing literature systematically. The goal of this study was to expand knowledge of the development tools for PHN competencies.

## 3 METHOD

This study utilised a systematic literature review. A systematic literature review is a piece of research in its own right that, by its nature, is able to address much broader questions than single empirical studies ever can (e.g. uncovering connections among many empirical findings) (Baumeister, 2013). Initially, the researchers used multiple databases of published literature, retrieved from ProQuest®, which appeared within the period 2013 to 2017. From the key search terms of *development instrument* and *PHN competency*, 50 articles were found in which the text or abstracts were written in English. The abstracts and full text of these articles were reviewed by the researcher (see Figure 1) and seven articles were selected according to the criteria for inclusion and exclusion. The inclusion criteria were: (1) originally published in English; (2) peer reviewed in international journals of nursing; (3) specifically focused on projects to develop and validate instruments of PHN competency. The exclusion criteria were: (1) use of unidimensional tools; (2) unclear tool development procedures; (3) lack of reliability or validity testing; (4) publications more than five years old.

| | |
|---|---|
| Identification → | Records identified through database searching (n = 22,055) |
| | ↓ |
| Screening → | Records after duplicates removed (n = 640) |
| | ↓ |
| | Records screened (n = 640) |
| | ↓ |
| | Records excluded (n = 21,415) |
| | ↓ |
| Eligibility → | Full-text articles assessed for eligibility (n = 50) |
| | ↓ |
| | Full-text articles excluded with reasons (n = 583) |
| | ↓ |
| Included → | Criteria exclusion<br>• use of unidimensional tools (n = 10)<br>• unclear tool development procedures (n = 8)<br>• lack of reliability or validity testing (n = 8)<br>• publications more than five years old (n = 17) |
| | ↓ |
| | Studies included in qualitative synthesis (n = 2) |
| | ↓ |
| | Studies included in quantitative synthesis (meta-analysis) (n = 5) |

Figure 1. Flow chart of study selection.

## 4 RESULTS

A summary of the articles reviewed will be explained, focusing on self-reporting tool development in PHN competency. Two themes emerged from the seven articles that met the inclusion and exclusion criteria, namely, the process of developing the tool, and the PHN competencies.

### 4.1 *Tool development process*

The process of developing a tool means showing the stages by which the tool is developed, as shown in Figure 2. There are two methods for developing a tool: first, by starting from an existing literature review; second, by gathering data through qualitative methods. Asahara et al. (2013), Ishii and Matsuda (2014), Duru et al. (2015), Gattinger et al. (2016), and Wu et al. (2016) developed specific items of instruments based on existing literature reviews and previous study. After creating a blueprint containing an item pool with response scale, the content analysis was done using a panel of experts. Following development of the tool item, a pilot test and psychometric properties were pursued to validate the instrument and make it reliable.

The second method of developing a tool is gathering data through qualitative methods. An exploratory approach using open-ended qualitative surveys and interviews can be through either or both of in-depth interviews and focus group discussions (Kitreerawutiwong et al., 2015; Heydari et al., 2016). The content validity was established through evaluation by experts. After the pilot testing was conducted, the psychometric testing was undertaken.

The process of validation of the instrument from seven studies was through content and construct validity by experts (Asahara et al., 2013; Ishii & Matsuda, 2014; Kitreerawutiwong et al., 2015; Duru et al., 2015; Heydari et al., 2016; Gattinger et al., 2016; Wu et al., 2016). They also calculated exploratory factor analysis and confirmatory factor analysis to validate their instruments. The reliability of studies used Cronbach's alpha.

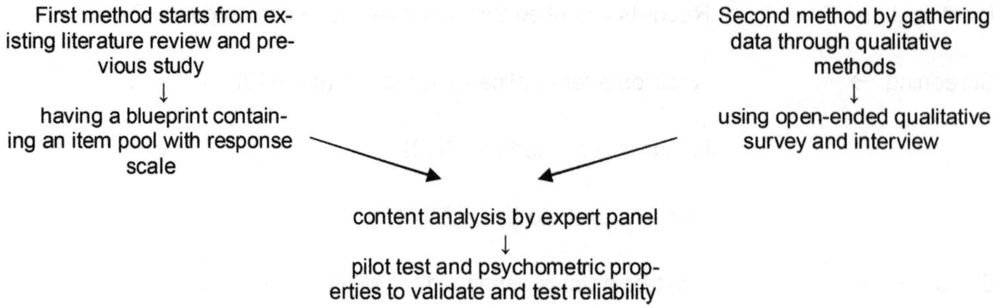
First method starts from existing literature review and previous study
having a blueprint containing an item pool with response scale
Second method by gathering data through qualitative methods
using open-ended qualitative survey and interview
content analysis by expert panel
pilot test and psychometric properties to validate and test reliability

Figure 2. Process of tool development.

### 4.2 *Public health nursing competencies*

PHN competencies were identified through analysing the dimensions of developed tools in literature reviews with regard to the ten essential public health services (see Table 1). The most essential service stressed in the developed tools is that of being assured of a competent public healthcare workforce (Asahara et al., 2013; Gattinger et al., 2016; Ishii & Matsuda, 2014; Kitreerawutiwong et al., 2015; Wu et al., 2016; Duru et al., 2015). The evaluation of personal and population-based health services is the second PHN competency to be explored (Asahara et al., 2013; Kitreerawutiwong et al., 2015; Wu et al., 2016; Duru et al., 2015); the third competency is collaborating or linking people to needed personal health services (Ishii & Matsuda, 2014; Kitreerawutiwong et al., 2015; Duru et al., 2015), and the fourth is the assessment of health status (Gattinger et al., 2016; Ishii & Matsuda, 2014). The fifth PHN competency is education for the patient and family (Asahara et al., 2013; Kitreerawutiwong et al., 2015), the sixth is interaction with the patient (Gattinger et al., 2016; Kitreerawutiwong et al., 2015), and the seventh is enforcement of regulation (Asahara et al., 2013; Wu et al., 2016). The last three services are investigation (Gattinger et al., 2016), policy making (Kitreerawutiwong et al., 2015), and defining the role through research (Ishii & Matsuda, 2014).

## 5 DISCUSSION

PHN researchers and practitioners in Minnesota developed an instrument measuring competency change after a population-focused continuing education series. Results showed that public health nurses had higher competency scores after the education series (Cross et al., 2006). Therefore, it is important to undertake a systematic review of these self-reporting tools. It is necessary to have a greater focus on the competency development of PHN. None of the assessment tools with acceptable validity and reliability were tested with public health nurses in Indonesia.

Each article showed strong validity. They were reviewed by a panel of experts and had strong results. The reliability of the articles was strong, with the exceptions of Heydari et al. (2016) with alpha 0.80, and Kitreerawutiwong et al. (2015) with alpha 0.7–0.88. Heydari et al. (2016) stated that because the scales comprised more than 15 items, the alpha value might have been exaggerated.

When developing tools, all of the articles were using mixed methods. Heydari et al. (2016) stated that the study was stronger than other studies when using a mixed-method design. The articles mentioned every phase of qualitative and quantitative procedures. Almost all articles started from a review of the existing literature, except Heydari et al. (2016) and Kitreerawutiwong et al. (2015).

In another study, Issel et al. (2006) developed and administered a competency self-assessment tool among PHN staff and the PHN faculty in Illinois, finding that nurses felt competent to practise in one of ten essential public health services, while the faculty felt competent in nine of these ten services. None, including from the faculty, felt adequately competent to teach any of the ten essential public health services. Gattinger et al. (2016) and Heydari et al. (2016) mentioned that including participants from different nursing care settings in a group was the strength of the study. More diversity in the participants used to validate the tools may provide more heterogeneity in the sample and more generalisability of the findings. All studies involved between five and eleven people on a panel, experts in PHN and developing tools, to validate their content and face.

## 6 CONCLUSION

Understanding the procedures used in developing tools is necessary when a researcher wants to make a proper tool to assess public health nursing competencies. It is important to have a panel of experts, a pilot test and psychometric properties to validate and the tools and test their reliability.

## REFERENCES


Asahara, K., Ono, W., Kobayashi, M., Omori, J. & Todome, H. (2013). Development and psychometric evaluation of the moral competence scale for home care nurses in Japan. *Journal of Nursing Measurement, New York, 21*(1), 43–54.

Baumeister, R.F. (2013). Writing a literature review. In M.J. Prinstein & M.D. Patterson (Eds.), *The portable mentor: Expert guide to a successful career in psychology* (2nd ed.; pp. 119–132). New York, NY: Springer Science + Business Media.

CDC. (2014). The public health system and the ten essential public health services. Atlanta, GA: Centers for Disease Control and Prevention. Retrieved from https://www.cdc.gov/stltpublichealth/publichealthservices/essentialhealthservices.html.

Cross, S., Block, D., Josten, L., Reckinger, D., Olson Keller, L., Strohschein, S. & Savik, K. (2006). Development of the public health nursing competency instrument. *Public Health Nursing, 23*(2), 108–114.

Duru, P., Örsal, Ö. & Karadağ, E. (2015). Development of an attitude scale for home care. *Research and Theory for Nursing Practice: An International Journal, 29*(4), 306–324.

Gattinger, H., Leoni-Kilpi, H., Hantikainen, V., Köpke, S., Ott, S. & Senn, B. (2016). Assessing nursing staff's competences in mobility support in nursing-home care: Development and psychometric testing of the Kinaesthetics Competence (KC) observation instrument. *BMC Nursing, 15*(1), 65.

Halcomb, E., Stephens, M., Bryce, J., Foley, E. & Ashley, C. (2016). Nursing competency standards in primary health care: An integrative review. *Journal of Clinical Nursing, 25*(9–10), 1193–1205.

Heydari, A., Kareshki, H. & Armat, M.R. (2016). How likely is it for a nurse student to become a competent nurse? A protocol for scale development and validation using a mixed methods study. *Acta Facultatis Medicae Naissensis, 33*(1), 49–61.

Hurd, A.R. & Buschbom, T. (2010). Competency development for chief executive officers in YMCAs. *Managing Leisure, 15*(1–2), 96–110.

Ishii, M. & Matsuda, N. (2014). Challenges of public health nurses in coordinating relationships: Scale development. *Social Behavior and Personality, 42*(6), 1029–1046.

Issel, L.M., Baldwin, K., Lyons, R.L. & Madamala, K. (2006). Self-reported competency of public health nurses and faculty in Illinois. *Public Health Nursing, 23*(2), 168–177.

Kitreerawutiwong, K., Sriruecha, C. & Laohasiriwong, W. (2015). Development of the competency scale for primary care managers in Thailand: Scale development. *BMC Family Practice, 16*(1), 174.

Kotrotsiou, S., Lavdaniti, M., Psychogiou, M., Paralikas, T., Papathanasiou, I. & Lahana, E. (2008). Community nurses' role as counsellors in primary health care. *International Journal of Caring Sciences, 1*(2), 92–98.

Lawson, T.G. (2014). Betty Neuman: Systems model. In M.R. Alligood (Ed.), *Nursing theorists and their work* (pp. 281–302). St. Louis, MO: Elsevier Mosby.

Philibin, C.A.N., Griffiths, C., Byrne, G., Horan, P., Brady, A. & Begley, C. (2010). The role of the public health nurse in a changing society. *Journal of Advanced Nursing, 66*(4), 743–752.

Prevost, C., Kpazai, G. & Attiklemé, K. (2015). Perceived importance of professional competencies for admission to the college of kinesiologists of Ontario. *International Journal of Kinesiology & Sports Science, 3*(2), 30.

Rowley, C. (2005). Health needs assessment. *Journal of Community Nursing*, *19*(6), 11–14.
Sakellari, E. (2012). Assessment of health needs: The health visiting contribution to public health. *International Journal of Caring Science*, 5(1), 19–26.
Swider, S., Krothe, J., Reyes, D. & Cravetz, M. (2013). The Quad Council practice competencies for public health nursing. *Public Health Nursing*, *30*(6), 519–536.
Wu, X.V., Enskär, K., Pua, L.H., Heng, D.G.N. & Wang, W. (2016). Development and psychometric testing of Holistic Clinical Assessment Tool (HCAT) for undergraduate nursing students. *BMC Medical Education*, *16*(1), 248.

*Strengthening Research Capacity and Disseminating New Findings in Nursing and Public Health – Malini et al. (Eds)*
*© 2018 Taylor & Francis Group, London, ISBN 978-1-138-50066-2*

# A systematic review: Implementation of reflective learning in nursing practice

J. Purwarini
*Sint Carolus School of Health Sciences, Jakarta, Indonesia*

J. Lorica
*St. Paul University Philippines, Cagayan, Philippines*

ABSTRACT: The aim of this study was to identify the use and effectiveness of reflective learning in nursing practice. This study used a systematic review of qualitative studies of reflective learning in nursing practice. Articles were searched for and identified from three databases (ProQuest, Cengage and EBSCO). Eleven studies were selected, with the total participants being 24 Register Nursing students, 315 nursing students, and 11 tutors in nursing. The methods used in these studies were interviews, reflective writing, focus group discussion and a combination of several methods. The three major themes were identified as the advantage of using reflective learning, barriers to reflection and the improvement due to the use of reflection in nursing education and nursing practice. The use of reflective learning in education and nursing practice is very effective in improving the quality of nursing students. All stakeholders would benefit from an awareness of the value of reflective learning, which can be implemented in patient care.

*Keywords*: Education, Nursing, Reflective learning

## 1 INTRODUCTION

The aim of nursing education is to prepare nursing students to have an appreciation and concern for human dignity during their duties in clinical service. Concern and caring are developed in order to enhance the ability of future nurses in rendering service, not only for the community, but also for the individual patients under their supervision. Each individual will be aware of the benefits of using reflective learning. Reflective learning is a learning process that requires a great deal of time and practice. It is a dynamic process that involves thinking through the issues ourselves, asking questions and seeking out related information in order to gain understanding. Reflection was found to develop students' self-awareness and self-confidence (Smith & Jack, 2005). Reflective learning has an advantage for the students' learning. Glaze (2002) discovered that critical reflection improved students' learning through a process of perspective transformation.

## 2 METHOD

The goal is to systematically review and synthesise the evidence from all of the published qualitative studies on reflective learning in supporting nursing practice. This integration of information from various qualitative studies may present a series of themes for the better understanding of the usage and effectiveness of reflective learning in nursing practice.

## 2.1 *Research strategy*

The literature is presented in a systematic review. The ProQuest, EBSCO, and Cengage libraries were searched from 1st January 2014 through to 31th March 2016, with the use of the terms reflective learning, reflective practice or reflective thinking, reflective writing or reflective journals, and nursing practice. These keywords were searched independently and in combination. In this first step of the search process, 177 articles were identified.

## 2.2 *Inclusion and exclusion criteria*

Published articles that met the following criteria were included:, the study should be in English; the participants are nursing students who are still studying or have recently graduated, Register Nurse (a nurse who has graduated from school nursing and has passed a national licensing exam) and the tutors. The studies must utilise qualitative methods on the research topic: reflective learning or reflection as a method of research, which includes reflective practice, reflective thinking, reflective writing and reflective journals. If duplicated studies were identified, only the most recent study was included. Only studies published as full text articles were involved in the review. Studies that were published only in abstract form were excluded.

## 2.3 *Data extraction*

The following general information was extracted from each article: language, author, year of publication, full text available, detail about the subject, and the method. The result of the systematic literature search is show in the following Figure 1:

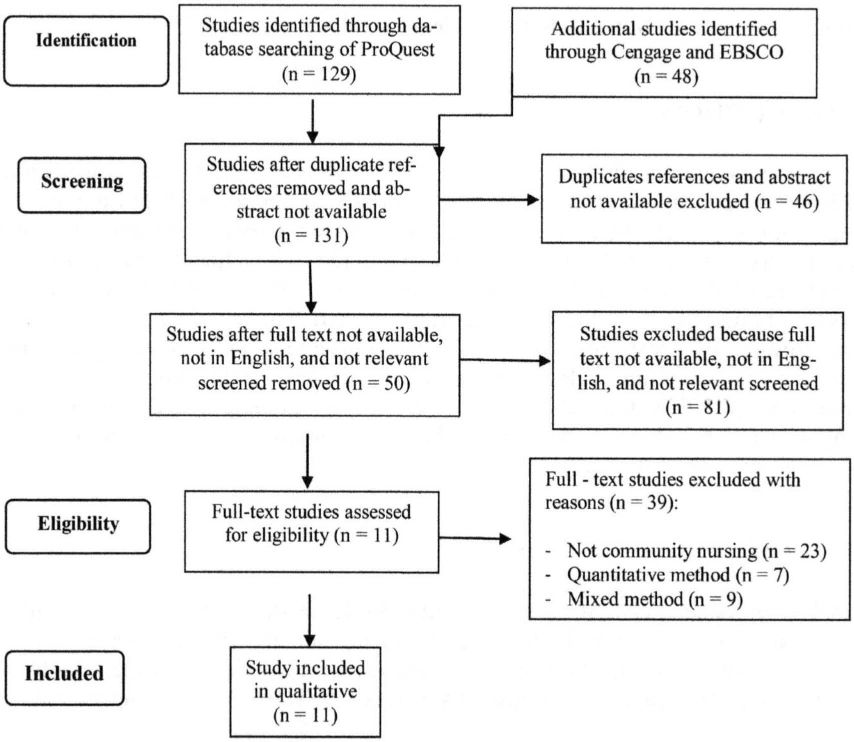


Figure 1. Results of the systematic literature search (Mother et al., 2009).

## 2.4 *Data analysis*

We included qualitative studies that used interviews, focus groups or observation on the use of reflective learning, and reflective writing in nursing practice. We excluded studies if they had duplicates references and the abstract was not available. Non-English articles were excluded in order to prevent a cultural and linguistic bias in translations. We excluded studies if they used mixed methods in their studies or reported only quantitative data. Studies that did not elicit data from the practice of nursing students within the scope of the hospital were also excluded.

# 3 RESULT

Out of the 50 studies that were examined, 11studies meet the inclusion criteria. The variables used for the study design was reflective learning, including reflective thinking, reflective writing, reflective case study, reflective interview, reflective action and reflective experiences. The participants involved in the 11studies consisted of 24 register nurse students, 315 nursing students, and 11 tutors in nursing. The purposes of most of the studies was to explore the perception and the effectiveness of reflective learning in enhancing the learning experience of nursing students. The same statement was also expressed by ter Maten-Speksnijder, et al. (2012) about the purpose of their study, which was to describe learning opportunities in the reflective case study, was used as an educational tool to inform future curriculum development. The other purpose of the studies was to explore how different didactic strategies support nursing students' experiences of learning during the first year of a reconstructed nursing curriculum (Westin et al., 2015; Willemse, 2015; Lister, 2012). Williams and Burke (2015) have a few objectives in their study: to analyse nursing students' stories and to understand how students develop a sense of being a nurse while pursuing a nursing degree. The use of reflective learning as a method in studies, could obtain data with regards to nurses' attitudes and goals in carrying out their profession. Many methods are used in reflective learning, such as reflective thinking, reflective writing, reflective case study, reflective experience, and reflective action.

## 3.1 *Advantage of using reflective learning*

The first theme is the advantage of using reflective learning. Some participants in the study stated that reflective learning increased their motivation in taking nursing education. The nursing students stated that they became more courageous in various situations. They trusted their ability and developed increased self-confidence. The students also believed in the importance of challenging their abilities and being themselves (Westin et al., 2015). The awareness of caring for the whole patient increased with the use of reflective learning (Westin et al., 2015; Williams & Burke, 2015). Several participants included the word "awesome"- as they reflected on learning experiences that caused them to "feel like a nurse" (Lister, 2012, p.111). Students have reflective ability (Silvia et al., 2012; ter Maten-Speksnijder et al., 2012), and student preconceptions of reflection, based on previous experience, shaped their perception of the benefits of the use of reflection in their current course. All students and tutors believed that reflection improved nursing practice and one tutor included benefits to future practice (Stirling, 2015).

The discussions suggest that reflective learning and the use of stories about the experience of giving and receiving care can contribute to the development of the knowledge, skills and confidence that enable student nurses to provide compassionate relationship—centred care within the practice (Adamson & Dewar, 2015). Positive learning experiences included access to a variety of clinical cases and information, the creation of a learning platform and the availability of educators and peers to answer questions (Willemse, 2015). Reflective experience provides benefits to participants, as mainly described incidents they had experienced; for example, patient data, doctor action, nurse action, treatment, communication, experience, and outcome. This experience will increase the confidence of students nurses (De Swardt et al., 2012).

### 3.2 *Barriers to reflection*

The second theme was barriers to reflection. Participants described faculty personalities and interactions as having very personal and significant impacts on their learning. It was reported that negative experiences were reported during reflective sessions due to the students taking advantage of each other and verbally the faculty members who did not like it if the student spoke frankly. This would result in a "tense classroom", which "made it hard to concentrate and learn" (Lister, 2012, p.102). All students and tutors thought that previous negative experiences of reflection impacted upon the students' perception of reflection. Too much reflection in post registered nursing education was reported to be burdensome (Stirling, 2015). In addition, it was also noted that study participants found themselves in a place full of disputes, which in some cases threatened the relationship with the student/instructor. Furthermore, reflective writing that was written by the participants, which described the instability in the clinical setting, was perceived as a source of stress for the participants and seemed to threaten the student-instructor relationship by decreasing tolerance and producing unfair expectations (Shahsavari et al., 2013).

### 3.3 *Improvements in the use of reflection in nursing education and nursing practice*

The last theme is the improvement in the use of reflection in nursing education and nursing practice. Stirling (2015) argued in his research that all students believed that reflection would be more beneficial if the students were given time to reflect on something they identified as important. Moreover, all students and tutors expressed the need to explore different models of reflection in order to maximise effectiveness. Reflective learning helps students to gain insights into nursing and increase one's self-awareness (Westin et al., 2015). Some participants ask students with extensive health care experience to note problems with their earlier experiences that they did not previously consider. They discovered many ways to care for patients and that they needed to utilise new strategies and knowledge for each unique situation. Furthermore, the students reported that a clinical placement early in the first semester was valuable for their learning. They became involved in real patient situations and had opportunities to care for patients in the early stages of the programme. The participants expressed reflective experience through focus group discussion (Lister, 2012). Clinical experiences also helped to develop a professional identity. Some participants expressed their awareness of growth while they were in school. Studies conducted by Willemse (2015), used electronic reflections. As a result, participants reflected that the electronic reflection discussions created an online discussion trail that allowed them to go back to information and use it in preparation for assessments.

## 4 DISCUSSION

Learning is equivalent to "finding meaning" and always implies schematisation and embodies new experiences. Learning implies construction and the approval of an interpretation that determines action (Silvia et al., 2012). The reflective practice could be described as a deliberate cognitive and affective exploration of experiences with the purpose of learning from experiences (Chapman et al., 2009). Different methods can be used for reflection, for example, reflective diaries, journals or writing, reflective group discussions, reflective experience and guided reflection. Reflective learning is the process of internally examining and exploring an issue of concern, triggered by an experience, which creates and clarifies meaning in terms of self, and which results in a changed conceptual perspective (Boyd & Fales, 1983). This process is central to understanding the experiential learning process. Reflective learning is a practice that facilitates the exploration, examination, and understanding of feeling, thinking and learning. It is a thoughtful consideration of academic material, personal experiences, and interpersonal relationships. It is a form of internal inquiry that extends the relevance of theory and deepens their understanding of the practice of everyday life and work. It is necessary, for education and nursing practice, to assess and evaluate all their quality of care. The most important aspect of reflective learning is a process in which people can learn about themselves. Furthermore, students knew where they made mistakes and could understand why the error occurred; therefore the same mistake will not be repeated. They discover their abilities and improve their self-confidence. Moreover, several studies also showed an increase in the quality of graduated students. Students will learn to recognise themselves, their abilities and their strengths, and will recognise senior habits and patterns of work in which they participate.

## 5 CONCLUSION

In conclusion, the use of reflective learning in education and nursing practice is very effective in improving the quality of nursing students. It is recommended that reflective learning should be included as a teaching and learning strategy method to enhance theory and practice integration in nursing. All role players, such as nurse educators, professional nurses, mentors, and preceptors, would benefit from an awareness of the value of reflective learning, not only for patient care but also for nurses' self-development.

## REFERENCES


Adamson, E. & Dewar, B., (2015). Compassionate care: Student nurses' learning through reflection and the use of story. *Nurse Education in Practice*, *15*(3), 155–161.

Ashley, J. & Stamp, K., (2014). Learning to think like a nurse: The development of clinical judgment in nursing students. *Journal of Nursing Education*, *53*(9), 519–525.

Boyd, E. M., & Fales, A. W. (1983). Reflective learning: Key to learning from experience. *Journal of Humanistic Psychology, 23*(2), 99–117.

Chapman, N., Dempsey, S. E. & Warren-Forward, H. M., (2009). Workplace diaries promoting reflective practice in radiation therapy. *Radiography*, *15*(2), 166–170.

De Swardt, H. C., Du Toit, H. S. & Botha, A., (2011). Guided reflection as a tool to deal with the theory-practice gap in critical care nursing students. *Health SA Gesondheid*, *17*(1), 1–9.

Edelen, B. G. & Bell, A. A., (2011). The role of analogy-guided learning experiences in enhancing students' clinical decision-making skills. *Journal of Nursing Education*, *50*(8), 453–460.

Glaze, J. E. (2002). Stages in coming to terms with reflection: Student advanced nurse practitioners' perceptions of their reflective journeys. *Journal of Advanced Nursing*, *37*(3), 265–272.

Lister, D. J. A., (2012). *Perspectives of practicing registered nurses on the importance of learning experiences in nursing school* (Doctoral dissertation, Walden University, Minneapolis, MN).

Moher, D., Liberati, A., Tetzlaff, J., Altman, D. G. & Prisma Group,. (2009). Preferred reporting items for systematic reviews and meta-analyses: The PRISMA statement. *PLoS Medicine*, *6*(7), p.e1000097.

Shahsavari, H., Yekta, Z. P., Houser, M. L. & Ghiyasvandian, S., (2013). Perceived clinical constraints in the nurse student–instructor interactions: A qualitative study. *Nurse Education in Practice*, *13*(6), 546–552.

Silvia, B., Valerio, D. & Lorenza, G., (2012). The reflective journal: A tool for enhancing experience-based learning in nursing students in clinical practice. *Journal of Nursing Education and Practice*, *3*(3), 102.

Smith, A. & Jack, K., (2005). Reflective practice: A meaningful task for students. *Nursing Standard*, *19*(26), 33–37.

Stirling, L., (2015). Students' and tutors' perceptions of the use of reflection in post-registration nurse education. *Community Practitioner*, *88*(4), 38–41.

ter Maten-Speksnijder, A. J., Grypdonck, M. H., Pool, A. & Streumer, J. N., (2012). Learning opportunities in case studies for becoming a reflective nurse practitioner. *Journal of Nursing Education*, *51*(10), 563–569.

Westin, L., Sundler, A. J. & Berglund, M., (2015). Students' experiences of learning in relation to didactic strategies during the first year of a nursing programme: A qualitative study. *BMC Medical Education*, *15*(1), 49.

Willemse, J. J., (2015). Undergraduate nurses reflections on WhatsApp use in improving primary health care education. *Curationis*, *38*(2), 1–7.

Williams, M. G. & Burke, L. L., (2015). Doing learning knowing speaking: How beginning nursing students develop their identity as nurses. *Nursing Education Perspectives*, *36*(1), 50–52.

*Strengthening Research Capacity and Disseminating New Findings in Nursing and Public Health – Malini et al. (Eds)*
*© 2018 Taylor & Francis Group, London, ISBN 978-1-138-50066-2*

# Determinants of nursing licensure examination performance: A literature review

Y. Siswadi
*Faculty of Nursing, Universitas Pelita Harapan, Jakarta*

A. Talosig
*School of Health Sciences, St. Paul University Philippines, Tuguegarao, Philippines*

ABSTRACT: Implementation of the Indonesian Nursing National Competency Test is facing several challenges. These issues have been a significant concern for every nursing school. The purpose of this study was to systematically review the existing literature to identify determinants or contributing factors in NLE (National Licensure Examination) performance. A multiple database search was used and out of 1,735 articles, 19 articles were reviewed which met both inclusion and exclusion criteria. The systematic review yielded two emerging themes in determining contributing factors to NLE performance namely: academic factors (cumulative grade point average, science subject scores, nursing subjects scores) and non-academic factors. Understanding the contributing factors to NLE performance is important not only for the nursing institutions but also for students and the faculty. The findings of this study provide deeper insights; contributing to the repository of knowledge of determinant or risk factors as bases for programme development for the success in the NLE performance.

## 1 INTRODUCTION

The implementation of the Indonesian Nursing National Competency Test (INNCT) has been facing several challenges, such as facilities and technical problems (internet connection, electricity, etc.), pass rate, and pass grade. The pass rate of period I/2015 was 45.45% and period II/2015 was 53.61% (Ristekdikti, 2016). These issues have been a significant concern for every nursing school, because they reflect the quality of the teaching process in the schools. There have been many studies conducted into National Licensure Examination (NLE) performance. This study aims to explore the determinants of NLE performance. The results could serve as database in preparing nurse graduates to perform better on the NLE.

## 2 METHOD

This systematic review study followed the Preferred Reporting Items for Systematic review and Meta-Analysis (PRISMA) protocol (See Figure 1). This involved conducting a systematic search of the literature using full electronic databases, including ProQuest®, OVID®, EBSCO®, JSTOR and PubMed® and a manual search of literature that appeared in the period from 2006 to March 2017. Inclusion criteria for this study included: research articles were written only in English, written within a ten-year period (2006–March 2017), peer-reviewed, has free full text, is published in a scholarly journal, and employed quantitative studies. The risks of bias include: the various published sources; the selection process used which consist of only five databases; and heterogeneity of the licensure examination or methods of the studies.

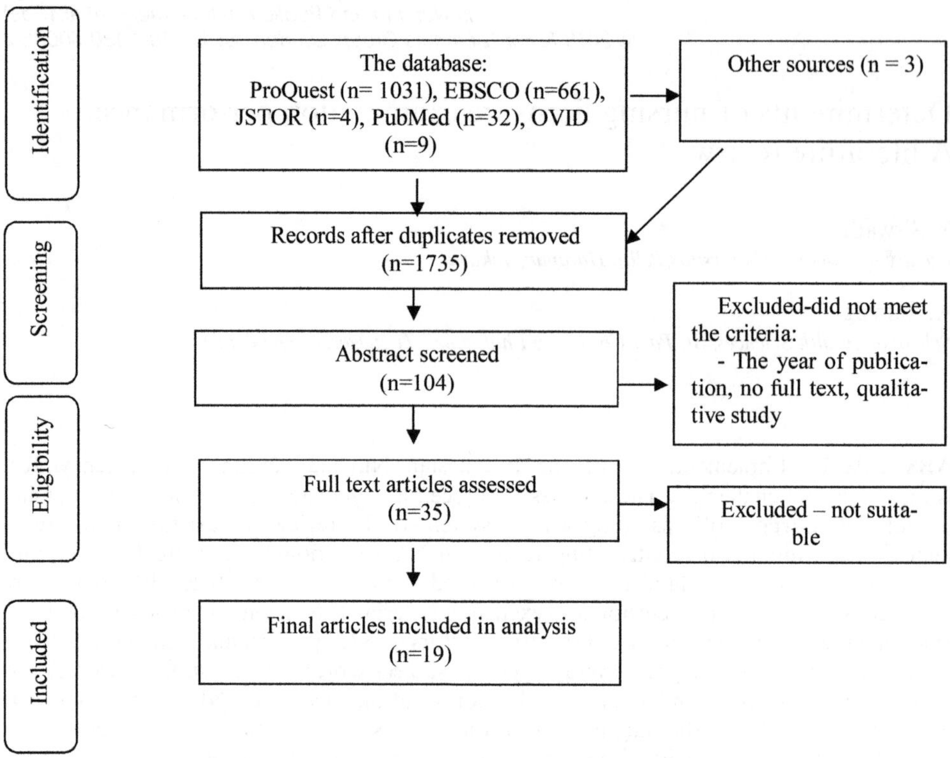


Figure 1. Flow diagram of reviewing process.

## 3 RESULTS

### 3.1 *Academic factors*

*Cumulative Grade Point Average (CGPA)*

The CGPA was considered as a significant predictor for NLE performance in several studies. Amankwaa et al. (2015) stressed that there was a strong association between CGPA and the NLE performance. Penprase et al. (2013) believed that overall GPA was increasing the probability of the pass rate. The higher the GPA, the higher is the possibility of passing the NCLEX-RN (National Council Licensure Examination for Registered Nurses) examination. Participants who were successful on the licensure examination at the first attempt had a 0.3 higher mean nursing GPA than those students who were unsuccessful (Gilmore, 2008). GPA demonstrated moderate positive correlations with NCLEX-RN success (Foley, 2016). The CGPA was found to have a significant relationship to, and could be predictive of, first-time success on the NCLEX-RN for graduates of the nursing programme (Reeve, 2014). A significant relationship was identified between the two variables ($r = 0.180$, $p < 0.01$) s GPA and the NCLEX-RN passing rate (Outlaw et al., 2013). Simon et al. (2013) argued that the GPA is a predictor of NLE score. In contrast, Ukpabi (2008) contended that there was no significant correlation between NCLEX-RN and GPA ($p = 0.676$).

### 3.2 *Science subject scores*

Breckenridge et al. (2012) stated that the best single predictor of NCLEX-RN passing rate was the science GPA, followed by the undergraduate GPA prior to the nursing major. Simon et al. (2013) supported that student performances in biology and chemistry courses are linearly associated with NLN (National League for Nursing)-readiness scores. Shirrell,

(2008)believed that the critical thinking score is predictive of success on the NCLEX-RN (F = 7.987, p 0.0001). Romeo (2013) maintained that the assessment test composite score (p = 0.013) and the critical thinking composite score (p = 0.008) were statistically significant as predictors of passing the NCLEX-RN for the first time. However, different results were found by McGahee et al., (2010) who explained there was no significant correlation between several variables such as science GPA, fundamentals of nursing, health assessment and pathophysiology. But, there were significant interactions that indicated that certain combined variables such as between the science GPA: fundamentals ($p = 0.002$), science GPA: health assessment ($p = 0.04$), science GPA: pathophysiology ($p = 0.02$) could be determinants for NLE performance.

### 3.3 *Nursing subjects scores*

Breckenridge et al. (2012) established that the undergraduate GPA prior to the nursing major is a significant predictor of passing the NCLEX-RN. Schooley & Kuhn (2013) recognised that the final course grade was significantly predictive of the HESI test ($p<0.01$). McGahee et al. (2010) explained that the main effects that were most significantly related to NCLEX-RN success were the RN (Registered Nurse) assessment test and theoretical foundations and pathophysiology. Penprase et al. (2013) showed several variables were significant contributors to predicting success on NCLEX-RN such as Comprehensive Adult Nursing I ($p = 0.004$) and pathophysiology. Abbott et al. (2008) found that there are statistically significant findings on the senior complex care grades and NCLEX-RN (p = 0.02). Simon et al. (2013) concluded that all nursing courses independently predict NLN-readiness scores. Romeo (2013) showed that the nursing GPA (p < 0.001) was the most powerful predictor of the first-time NCLEX-RN pass rate. In addition, Leon (2016) indicated there was low correlation (p < 0.001) between NLE performance and academic performance. Nacos-Burds, (2010) found that the practical nursing core GPA was found to be a significant predictor of NCLEX-RN success. In line with the previous mentioned study, McGahee et al. (2010) point out that there was no significant correlation between several variables such as health assessment and pathophysiology and NLE performance.

### 3.4 *Non-academic factors*

Amankwaa et al. (2015) found that there was no statistically significant correlation between sociodemographics such as: gender ($p = 0.288$), age ($p = 0.180$), Christian religion ($p = 0.210$), description of home community ($p = 0.919$), mother's education ($p = 0.917$), and father's education ($p = 0.796$) and performance in licensure examination. Breckenridge et al. (2012) identified that the best predictor of NCLEX-RN pass rate was family income. Benefiel ( 2011) addressed that gender, ethnicity, and age shown did not have a significant relationship with NCLEX-RN performance. Another study by Whitehead (2016) recognised that the gender and age of the students were not significant predictors of NCLEC-RN performance. Gutierrez (2016) presented that there was significant correlation between the school accreditation status and board performance. Simon et al. (2013) suggest that the transferred students and GPA were significantly predictor of the NLN (National League for Nursing) score Ukpabi (2008) discovered that out of 18 predictor variables in the Assessment Technologies Institute (ATI), only 11 were significant in predicting pass rates of the NCLEX-RN such as: critical thinking, Test of Essential Academic Skills (TEAS), reading, maths, English, mental, pharmacology, fundamental, National League Nursing (NLN) Adult1, NLN adult2, and NLN Paediatric.

## 4 DISCUSSION

Understanding the contributing factors to NLE performance is important not only for the nursing institutions but also students, and the faculty Passing the licensure examination is required for registration and practice as a nurse. The licensure examination is designed to identify candidates who possess the theoretical knowledge to practice as an entry-level nurse. The findings from this study indicate that the academic factors are the most researched and provide strong evidence to predict success on the NLE. Student's GPA, science score and nursing subjects have a stronger value compare to others (Amankwa et al., 2015; Penprase et al., 2013; Simon et al., 2013; Gilmore, 2008; Reeve, 2014; Outlaw et al., 2013). It is essential a lecturer can monitor students' progress and support them every semester to achieve high grades in every subject. The students could think to attend a special programme to improve their knowledge.

Non-academic factors such as gender, age, religion, parent's education background and family income were mentioned in the studies but not all have significant correlation with the NLE performance. Previous studies did not support the relationship between age, gender, religion, parent's education background and NLE performance (Amankwa et al., 2015;Benefiel, (2011) Whitehead, 2016). This finding is inconsistent with the findings from other researchers who assert that older students perform better (Simon et al., 2013). Accreditation status and level have a significant relationship with NLE performance. Accreditation is the process by which schools are evaluated based on specific standards to ensure the quality of the learning process. It can stimulate institutions to achieve maximum standards and to identify schools whose competence in a particular field warrants public and professional recognition (Gutierrez, 2016). Accreditation is a process that is recognised worldwide as an external quality assurance.

## 5 CONCLUSION

There is much work to be done to advance the identification of determinants of NLE performance. Several significant predictors were identified as academic and non-academic factors. Testing this finding could be a great input as basis for reflection and improvement. Nursing schools should think in terms of updating curricula, teaching styles, or special programmes to support student performance.

## REFERENCES


Abbott, A.A., Schwartz, M.M., Hercinger, M., Miller, C.L. & Foyt, M.E. (2008). Predictors of success on national council licensure examination for registered nurses for accelerated baccalaureate nursing graduates. *Nurse Educator*, *33*(1), 5–6.

Abbott, A. A., Schwartz, M. M., Hercinger, M., Miller, C. L., & Foyt, M. E. (2008). Predictors of Success on National Council Licensure Examination for Registered Nurses for Accelerated Baccalaureate Nursing Graduates. *Nurse Educator*, *33*(1), 5–6. https://doi.org/10.1097/01.NCN.0000336453.62659.06.

Amankwaa, I., Agyemang-Dankwah, A. & Boateng, D. (2015). Previous education, sociodemographic characteristics, and nursing cumulative grade point average as predictors of success in nursing licensure xaminations. *Nursing Research and Practice*, *2015*, 682479.

Amankwaa, I., Agyemang-Dankwah, A., & Boateng, D. (2015). Previous Education, Sociodemographic Characteristics, and Nursing Cumulative Grade Point Average as Predictors of Success in Nursing Licensure Examinations. *Nursing Research and Practice*, *2015*, 682479. https://doi.org/10.1155/2015/682479.

Benefiel, D. (2011). Predictors of success and failure for Adn atudents on the NCLEX-RN. *ProQuest Dissertations and Theses*. California State University, Fresno. (UMI No. 3456526).

Benefiel, D. (2011). *Predictors of success and failure for adn studnts on the NCLEX-RN*. Retrieved from http://e-resources.perpusnas.go.id:2071/docview/871704955/fulltextPDF/1 A12DCFE328A434FPQ/7?accountid=25704.

Benefiel, D., & Diane. (2011). *Predictors of Success and Failure for Adn Students on the Nclex-Rn. ProQuest Dissertations and Theses*. California State University, Fresno. Retrieved from https://www.mendeley.com/research-papers/predictors-success-failure-adn-students-nclexrn/?utm_source=desktop&utm_medium=1.17.8&utm_campaign=open_catalog&userDocumentId=%7B79587fda-8dea-3363-a2dd-a4f268496114%7D.

Breckenridge, D.M., Wolf, Z.R. & Roszkowski, M.J. (2012). Risk assessment profile and strategies for success instrument: Determining prelicensure nursing students' risk for academic success. *Journal of Nursing Education, 51*, 160–166.
Breckenridge, D. M., Wolf, Z. R., & Roszkowski, M. J. (2012). Risk Assessment Profile and Strategies for Success Instrument: Determining Prelicensure Nursing Students' Risk for Academic Success. *Journal of Nursing Education, 51*, 160–166. https://doi.org/10.3928/01484834-20120113-03.
Foley, D.M. (2016). *Predicting student success: Factors influencing nclex-rn ® rates in an urban univer sity' s pre-licensure programs. ProQuest Dissertations and Theses*. Cleveland State University, Cleveland. (UMI No. 10115724).
Foley, D. M. (2016). *Predicting student success: factors influencing NCLEX-RN ® rates in an urban university's pre-licensure program. (dissertatation) Retrieved from ProQuest: 10115724.*
Gilmore, M. (2008). Predictors of success in associate degree nursing programs. *Teaching and Learning in Nursing, 3*(4), 121–124.
Gilmore, M. (2008). Predictors of success in associate degree nursing programs. *Teaching and Learning in Nursing, 3*(4), 121–124. https://doi.org/10.1016/j.teln.2008.04.004.
Gutierrez, N.P. (2016). Level of accreditation and board performance of the colleges of nursing in the national capital region. *International Education & Research Journal, 2*(5), 21–29.
Indonesian Ministry of Research and Technology of Higher Education. (2016). National competency test of health as a concrete step of quality assurance of higher education of health Retrieved 16 May 2017, from http://www.dikti.go.id.
Leon, D. (2016). Academic and licensure examination performances of BSN graduates: Bases for curriculum enhancement. *International Journal of Educational Policy Research and Review, 3*(4), 64–72.
Leon, D. (2016). Academic and licensure examination performances of BSN graduates: Bases for curriculum enhancement. *International Journal of Educational Policy Research and Review, 3*(4), 64–72. https://doi.org/10.15739/IJEPRR.16.009.
McGahee, T.W., Gramling, L. & Reid, R.F. (2010). NCLEX-RN success: Are there predictors? *Southern Online Journal of Nursing Research, 10*(4), 1–9.
McGahee, T. W., Gramling, L., & Reid, R. F. (2010). NCLEX-RN Success: Are There Predictors? *Southern Online Journal of Nursing Research, 10*(4), 1–9.
Nacos-Burds, K. J. (2010). A comparative analysis of demographic and academic characteristicsand NCLEX-RN passing among urban and rural campus students in a Midwest Associate Degree Nursing Program. *Graduate Theses and Dissertations.* Iowa State University, Ames. (Paper 11639). Retrieved from http://lib.dr.iastate.edu/etd.
Nacos-Burds, K. J. (2010). *A comparative analysis of demographic and academic characteristicsand NCLEX-RN passing among urban and rural campus students in a Midwest Associate Degree Nursing Program. Dissertation. Paper 11639.* Retrieved from http://lib.dr.iastate.edu/etd.
Outlaw, K., Burn, D., Rushing, D., Spurlock, A., Dunn, C., Bazzell, J. & Cleveland, K. (2013). Academic & non-academic variables impacting BSN students who are unsuccessful on onitial NCLEX-RN. *International Journal of Nursing and Health Care, 1*(1), 143–147.
Outlaw, K., Burn, D., Rushing, D., Spurlock, A., Dunn, C., Bazzell, J., Cleveland, K. (2013). Academic & Non-Academic Variables Impacting BSN Students Who are Unsuccessful on Initial NCLEX-RN. *International Journal of Nursing and Health Care, 1*(1), 143–147. https://doi.org/10.5176/2345-718X.
Penprase, B.B., Harris, M. & Qu, X. (2013). Academic success: Which factors contribute signifycantly to NCLEX-RN success for ASDN students?. *Journal of Nursing Education and Practice, 3*(7), 1–8.
Penprase, B. B., Harris, M., & Qu, X. (2013). Academic success: Which factors contribute signifycantly to NCLEX-RN success for ASDN students?, *3*(7), 1–8. https://doi.org/10.5430/jnep.v3n7p1.
Reeve, I. (2014). Predictors of first-time success on the NCLEX-RN for graduates graduates of a baccalaureate nursing program. *ProQuest Dissertations and These.* University of South Dakota, Vermillion. (UMI No. 3640298)
Reeve, I. (2014). Predictors of first-time success on the NCLEX-RN for graduates of baccalaurate nursing program. *(Dissertation) Retrieved from ProQuest:3640298*, (August).
Ristekdikti. (2016). Implementasi Uji Kompetensi Nasional bidang Kesehatan sebagai Langkah Konkrit Penjaminan Mutu Pendidikan Tinggi Kesehatan—Ristekdikti. Retrieved May 16, 2017, from http://www.dikti.go.id/implementasi-uji-kompetensi-nasional-bidang-kesehatan-sebagai-langkah-konkrit-penjaminan-mutu-pendidikan-tinggi-kesehatan/
Romeo, E.M. (2013). The predictive ability of critical thinking, nursing GPA, and SAT scores on first-time NCLEX-RN performance.pdf. *Nursing Education Perspectives, 34*(4), 248–253.

Romeo, E. M. (2013). The predictive ability of critical thinking, nursing GPA, and SAT scores on first-time NCLEX-RN performance.pdf. *Nursing Education Perspectives*, *34*(4), 248–253. https://doi.org/10.5480/1536-5026-34.4.248.

Schooley, A. & Kuhn, J.R.D. (2013). Early indicators of NCLEX-RN performance. *The Journal of Nursing Education*, *52*(9), 539–42.

Schooley, A., & Kuhn, J. R. D. (2013). Early Indicators of NCLEX-RN performance. *The Journal of Nursing Education*, *52*(9), 539–542. https://doi.org/10.3928/01484834-20130819-08.

Shirrell, D. (2008). Critical thinking as a predictor of success in an associate degree nursing program. *Teaching and Learning in Nursing*, *3*, 131–136.

Shirrell, D. (2008). Critical thinking as a predictor of success in an associate degree nursing program. *Teaching and Learning in Nursing*, *3*, 131–136. https://doi.org/10.1016/j.teln.2008.05.001.

Simon, E.B., McGinniss, S.H. & Krauss, B.J. (2013). Predictor variables for NCLEX-RN readiness exam performance. *Nursing Education Research*, *34*(1), 18–24.

Simon, E.B., McGinniss, S, H., & Krauss, B, J. (2013). Predictor Variables for NCLEX-RN Readiness Exam Performance. *Nursing Education Research*, *34*(1), 18–24.

Ukpabi, C.V. (2008). Predictors of successful nursing education outcomes: A study of the orth Carolina Central University's nursing program. *Educational Research Quarterly*, *32*(2), 30–40.

Ukpabi, C. V. (2008). Predictors of Successful Nursing Education Outcomes: A Study of the North Carolina Central University's Nursing Program. *Educational Research Quarterly*, *32*(2), 30–40.

Whitehead, C.D. (2016). Predicting national council licensure examination for registered nurses performance. *ProQuest Dissertations and These. Northcentral University, Arizona.* (UMI No. 10113822).

Whitehead, C. D. (2016). Predicting National Council Licensure Examination for Registered Nurses Performance. *Dissertation. Retrieved from ProQuest: 10113822*. Retrieved from Retrieved from http://e-resources.perpusnas.go.id:2071/docview/1802335718/fulltextPDF/1A12DCFE328A434FPQ/1?accountid=25704.

*Pediatric health*

*Strengthening Research Capacity and Disseminating New Findings in Nursing and Public Health – Malini et al. (Eds)*
 *ISBN 978-1-138-50066-2*

# The effect of a kaleidoscope on pain relief during a venepuncture procedure in children in Padang, Indonesia

D. Novrianda, R. Fatmadona & L.B. Safira
*Faculty of Nursing, Andalas University, Padang, West Sumatra, Indonesia*

ABSTRACT: Various nursing actions and treatment procedures in hospital often cause pain in sick children. The purpose of this study was to analyse the effect of distraction with a kaleidoscope on pain and vital signs. A quasi-experimental investigation was developed in the Emergency Department of Dr M. Djamil Hospital Padang, West Sumatra Province, Indonesia. A consecutive sampling technique was utilised with 20 children being selected. The children were randomly divided into two groups. The mean pain score in the experimental group was 0.80 and in the control group was 3.40, as measured using the Wong-Baker Faces Pain Rating Scale (WBF-PRS). The Mann–Whitney $U$ test indicated that there were differences in both groups ($p = 0.038$), with kaleidoscope therapy decreasing diastolic blood pressure ($p = 0.018$) and respiratory rate ($p = 0.024$). The kaleidoscope was a very effective distraction during venepuncture to minimise pain and physiological response. It is suggested that nurses can employ distraction as a routine care element, especially when applying medical procedures.

*Keywords*: pain, distraction, kaleidoscope, vital signs, veneipuncture

## 1 INTRODUCTION

Pain is a subjective experience that commonly occurs in children and can be caused by actual or potential tissue damage. Pain in children is difficult to accurately identify. Consequently, pain management can be ineffective and cause negative impacts such as increased intensity, frequency, duration or degree of pain-related damage to children's bodies (Truba & Hoyle, 2014). Pain can have a detrimental influence for physical, emotional, behavioural, cognitive and psychological aspects (Czarnecki et al., 2011; Taddio et al., 2010). Adverse effects can occur such as fear, anxiety and refusal for subsequent procedures (Czarnecki et al., 2011; Taddio et al., 2010), syringe phobia (Taddio et al., 2010), aggressive behaviour and distrust of health care workers (Czarnecki et al., 2011). In addition, physical aspects can affect body systems like cardiopulmonary function, metabolism and the immune system (Czarnecki et al., 2011).

Venepuncture is one of the minor medical procedures that are performed and cause acute pain in children (Sikorova & Hrazdilova, 2011). Venepuncture is the second most common procedure that can cause moderate to severe pain in children (Stevens et al., 2011), and Hartling et al. (2013) stated that venepuncture and intravenous infusion are the most common procedures in the emergency department.

Professional nurses should understand the importance of pain management (Wong et al., 2012). Pain management is divided into two approaches, namely pharmacological and non-pharmacological (Taddio et al., 2010). Distraction is one of the non-pharmacological interventions that distracts children's attention from painful stimuli (El-Gawad & Elsayed, 2015), a cognitive-behavioural approach to decreasing pain during invasive procedures on children in the emergency department (Wente, 2013). Distraction is a nursing intervention that is easy, inexpensive, and effective (Bagheriyan et al., 2012), and adjusts vital signs in the physiological response of pain (El-Gawad & Elsayed, 2015; Kiani et al., 2013).

Many experimental studies on distraction reduce pain significantly, including the use of kaleidoscopes (Birnie et al., 2014). A kaleidoscope is a toy that attracts children's attention when they look into it, so that they do not focus on the pain of invasive procedures (Tüfekci et al., 2009). Canbulat et al. (2014) stated that using a kaleidoscope may result in lower pain scores in school-age children during venepuncture. In Indonesia, the study of distraction with a kaleidoscope as a form of pain management has not yet been undertaken. Thus, the purpose of this study was to determine the effect of distraction using a kaleidoscope as a form of pain relief for children undergoing venepuncture.

## 2 METHODS

The method was a quasi-experimental, aiming to provide an overview of each variable and determine the effect of distraction using a kaleidoscope on reducing pain scores and changing the vital signs of children undergoing venepuncture. The population involved children who visited the emergency department. The sampling method was consecutive, yielding a total of 20 samples.

The inclusion criteria were: 1) aged 6–11 years old; 2) undergoing venepuncture; 3) able to communicate verbally and non-verbally; 4) parents are willing to be respondents. Exclusion criteria were: 1) in a critical condition; 2) uncooperative parents. The research was conducted from 11 August to 3 October 2016 in the Emergency Department of Dr M. Djamil Hospital, Padang. Data collection was conducted using the Wong-Baker Faces Pain Rating Scale (WBF-PRS), wrist blood pressure and a stopwatch.

This study obtained ethical clearance from the Ethical Commission of the Faculty of Medicine, University of Andalas. The parents provided written informed consent prior to data collection. Confidentiality was guaranteed regarding information and the children's identity. The statistical test applied to prove the hypothesis was the Mann–Whitney *U* test, as the data were not normally distributed ($p < 0.05$).

## 3 RESULTS

Table 1 illustrates that most of the children in both groups were nine years old. In terms of gender, the control group was dominated by male children, while the experimental group had an equal number of both sexes. Regarding previous experience of venepuncture, most children in the control group had a previous history. Meanwhile, more than half of the children in the

Table 1. Respondent characteristics by age, gender and experience of blood sampling.

| | Control | Experimental |
|---|---|---|
| Characteristic | *f* (%) | *f* (%) |
| Age (years): | | |
| 6 | 0 (0) | 2 (20) |
| 7 | 2 (20) | 2 (20) |
| 8 | 2 (20) | 1 (10) |
| 9 | 4 (40) | 3 (30) |
| 10 | 1 (10) | 0 (0) |
| 11 | 1 (10) | 2 (20) |
| Gender: | | |
| Male | 7 (70) | 5 (50) |
| Female | 3 (30) | 5 (50) |
| Previous venepuncture: | | |
| No | 3 (30) | 4 (40) |
| Yes | 7 (30) | 6 (60) |

Table 2. Comparison of pain level felt by children during the procedure.

| Variable | Group | Mean | Min | Max | *p* value |
|---|---|---|---|---|---|
| Pain score | Control | 3.40 | 0 | 4 | 0.038 |
| | Experimental | 0.80 | 0 | 8 | |

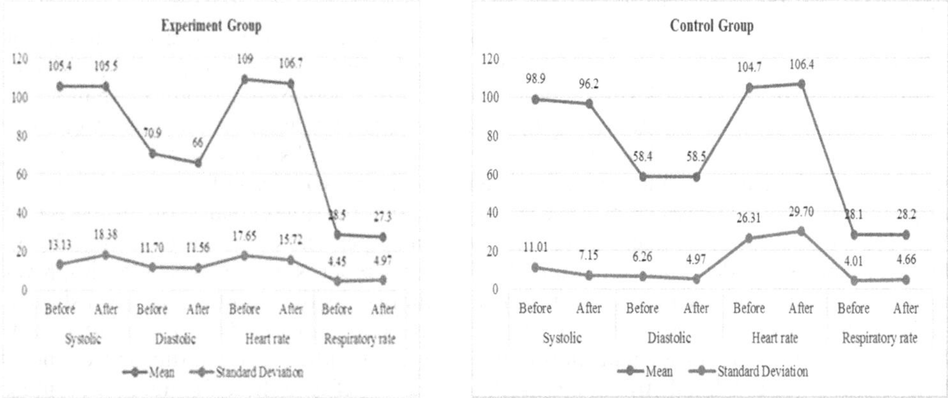


Figure 1. Comparison of vital signs before and after distraction during the procedure.

Table 3. Vital signs comparison between control and experimental groups after the procedure.

| Variable | Group | Mean | SD | *p* value |
|---|---|---|---|---|
| Systolic b.p. | Control | 96.20 | 7.146 | 0.153 |
| | Experimental | 105.50 | 18.380 | |
| Diastolic b.p. | Control | 58.50 | 4.972 | 0.076 |
| | Experimental | 66.00 | 11.557 | |
| Pulse rate | Control | 106.40 | 29.703 | 0.978 |
| | Experimental | 106.70 | 15.702 | |
| Respiratory rate | Control | 28.20 | 4.662 | 0.681 |
| | Experimental | 27.30 | 4.968 | |

experimental group had a history. Table 2 indicates that there was a significant difference in pain scores between the control and experimental groups, with a *p* value of 0.038.

From Figure 1, the *p* values of systolic blood pressure and pulse rate were 0.977 and 0.225, respectively. This shows that there was no significant difference in systolic blood pressure and pulse rate before and after administering distraction. Meanwhile, *p* values of diastolic blood pressure and respiratory rate were 0.018 and 0.024, respectively, which means that there was a significant difference before and after venepuncture using the kaleidoscope. Looking at the *p* values in the control group for systolic and diastolic blood pressures, pulse and respiratory rate, they were, respectively, 0.175, 0.614, 0.323 and 0.780. The figure shows that there was no significant difference before and after venepuncture.

As displayed in Table 3, all *p* values were above 0.05, indicating that there was no significant difference between the control and experimental groups after venepuncture.

## 4 DISCUSSION

The average pain score in the experimental group after kaleidoscope distraction was 0.80. The lowest value was 0 (no pain) and the highest was 4 (slightly more painful). Overall, most respondents did not feel any pain during venepuncture while using the kaleidoscope. However, the average pain score of the control group was 3.40. The lowest value was 0 and the highest was 8 (very painful). Overall, most respondents felt pain ranging from slightly painful to very painful and only a few children had no pain during venepuncture. The statistical analysis proved that there was a significant difference in pain scores between the experimental and control groups. The present findings seem to be consistent with other studies that found that a kaleidoscope significantly reduced pain (Karakaya & Gözen, 2015; Canbulat et al., 2014; George & Vetriselvi, 2015).

There are possible explanations for this result. The selection of a proper distraction for children is very important, so that their attention can be diverted from the painful procedure (Bagheriyan et al., 2012). A kaleidoscope as a visual distraction that produces shapes and colours can be an option (Canbulat et al., 2014). It is not only its shape and colour but also when the kaleidoscope is circling, it creates various colours and shapes such as beads which are uniquely adapted to the movement (Tüfekci et al., 2009). Children's focus on the kaleidoscope can distract them from painful procedures. Another possible explanation for this is that distraction can provide an analgesic effect by altering a person's emotions (Johnson, 2005). These can be changed by providing a distraction that reduces anxiety and then improves mood and motivation. As a result, pain can be minimised, as the attention has been shifted. In addition, distraction can also effectively reduce pain due to the child's willingness to use it. The shapes and colours inside a kaleidoscope encourage children to use it during treatment. Johnson (2005) noted that distraction is an effective and useful method if children are willing to use it.

Diastolic blood pressure before venepuncture was 70.90 and declined thereafter to 66.00, with a $p$ value of 0.018. This means that there was a significant difference in average diastolic blood pressure before and after venepuncture with distraction. A similar trend was found in respiratory rate, where an average of 28.50 was recorded before venepuncture and decreased thereafter to 27.30, with a $p$ value of 0.024. This result is in agreement with the findings of El-Gawad and Elsayed (2015), which reported a decrease of systolic and diastolic blood pressures, pulse and respiratory rate, with $p$ values, respectively, of 0.014, 0.023, 0.001 and 0.002 in the experimental group before and after the distraction. This means that distraction had a significant influence on vital signs.

What is surprising is that the kaleidoscope did not affect systolic blood pressure and pulse rate. In this study, there was a slight increase in systolic blood pressure. It seems that these results are possible due to the fact that some of respondents have problems with their kidneys. Kidney problems can lead to intravascular fluid changes that have an impact on stimulation of the sympathetic nervous system (Craven & Hirnle, 2009). Meanwhile, the pulse frequency slightly decreased. Distraction can help minimise pain and anxiety that impact on the vital signs. In line with Farrokhnia et al. (2011), distraction can reduce the chemical and physiological changes resulting from anxiety and discomfort due to invasive procedures. It will inactivate the autonomic nervous system, hence there is no increase in vital signs. Therefore, distraction is very efficient and considered for use in such invasive procedures.

In the control group, the current study found that the systolic and diastolic blood pressures went down after venepuncture, while the pulse and respiratory rate rose thereafter. To our surprise, no differences were found in vital signs before and after the procedure. This result is in agreement with the findings of McClellan et al. (2009), which revealed that the pulse rate did not show any significant differences before and after venepuncture. In contrast, Hosseini et al. (2016) reported that there were differences in vital signs before and after a bone marrow aspiration without distraction. A possible explanation for this might be that the pain felt by the individual in the control group may cause changes in vital signs. In accordance with Farrokhnia et al. (2011), the signal of pain is received by the hypothalamus, which then stimulates the sympathetic nervous system, causing an escalation in pulse rate and blood pressure and uplifts in intake of oxygen.

This study indicates that there was no difference in vital signs of both groups after the procedure. These results matched those observed in earlier studies. Hosseini et al. (2016) indicated that there was no significant difference in the vital signs after the procedure. Further, these findings supported the results of Hartling et al. (2013) who also found no significant difference in pulse rate during invasive procedures between a group receiving music therapy and a control group. A possible explanation for these results may be due to the activation of the autonomic nervous system, which is caused by pain, anxiety, exercise and changes in intravascular volume (Craven & Hirnle, 2009).

In this study, there were haemophilia patients in the control group. Haemophilia can cause a fluid volume deficiency. Lack of fluid volume can reduce a child's blood pressure (Craven & Hirnle, 2009). This may cause the blood pressure in the control group to be lower than in the experimental group. In terms of respiratory rate, the control group had a higher rate than the experimental group. This may happen due to the stress of hospitalisation for children (Craven & Hirnle, 2009). The limitations of this study include the number of samples that are lacking, as well as the diagnosis of diseases as a difference that may influence the vital signs.

## 5 CONCLUSION

This study concludes that distraction using a kaleidoscope can effectively reduce pain and vital signs adjustment. There was a significant difference in pain scores between the experimental and control groups. A kaleidoscope can be an option for nurses to minimise pain in children due to invasive measures such as venepuncture.

## REFERENCES


Bagheriyan, S., Borhani, F., Abbaszadeh, A., Miri, S., Mohsepour, M. & Zafarnia, N. (2012). Analgesic effect of regular breathing exercise with the aim of distraction during venipuncture in school-aged thalassemic children. *Iranian Journal of Pediatric Hematology Oncology*, *2*(3), 116–122.

Birnie, K.A., Noel, M., Parker, J.A., Chambers, C.T., Uman, L.S., Kisely, S.R. & McGrath, P.J. (2014). Systematic review and meta-analysis: Distraction and hypnosis for needle-related pain and distress in children and adolescents. *Journal of Pediatric Psychology Advances Access, 38*(8), 783–808.

Canbulat, N., Inal, S. & Sönmezer, H. (2014). Efficacy of distraction methods on procedural pain and anxiety by applying distraction cards and kaleidoscope in children. *Asian Nursing Research*, *8*(1), 23–28.

Craven, R.F. & Hirnle, C.J. (2009). *Fundamentals of nursing* (6th ed.). Philadelphia, PA: Wolters Kluwer.

Czarnecki, M.L., Turner, H.N., Collins, P.M., Doellman, D., Wrona, S. & Reynolds, J. (2011). Procedural pain management: A position statement with clinical practice recommendations. *Pain Management Nursing*, *12*(2), 95–111.

El-Gawad, S.M.E. & Elsayed, L.A. (2015). Effect of interactive distraction versus cutaneous stimulation for venipuncture pain relief in school age children. *Journal of Nursing Education and Practice*, *5*(4), 32–40.

Farrokhnia, M., Fathabadi, J. & Shahidi, S. (2011). Investigating the effects of cognitive interventions on reducing pain intensity and modifying heart rate and oxygen saturation level. *Journal of Jahrom University of Medical Sciences*, *9*(3), 26–33.

George, R.E. & Vetriselvi. (2015). Effect of kaleidoscope as distractor on pain. *Indian Journal of Surgical Nursing*, *4*(3), 77–83.

Hartling, L., Newton, A.S., Liang, Y., Jou, H., Hewson, K., Klassen, T. & Curtis, S. (2013). Music to reduce rain and distress in the pediatric emergency department: A randomized clinical trial. *JAMA Pediatric*, *167*(9), 826–835.

Hosseini, S.E., Moeninia, F. & Javadi, M. (2016). The effect of distraction method on bone marrow aspiration pain: A randomized clinical trial. *Journal of Biology and Today's World*, *5*(1), 20–24.

Johnson, M.H. (2005). How does distraction work in the management of pain. *Current Pain and Headache Reports*, *9*, 90–95.

Karakaya, A. & Gözen, D. (2015). The effect of distraction on pain level felt by school-age children during venipuncture procedure-randomized controlled trial. *Pain Management Nursing*, *17*(1), 47–53.

Kiani, M.A., Nagaphi, M., Jafaril, S.A., Mobarhan, M.A., Mohammadi, S., Saedi, M., ... Ferns, G. (2013). Effect of music on pain, anxiety, and vital signs of children during colonoscopy. *Life Science Journal*, *10*(12s), 31–33.

McClellan, C.B., Schatz, J.C., Mark, T.R., McKelvy, A., Puffer, E., Roberts, C. & Sweltzer, S.M. (2009). Criterion and convergent validity for four measures of pain in a pediatric sickle cell disease population. *The Clinical Journal of Pain*, *25*(2), 146–152.

Sikorova, L. & Hrazdilova, P. (2011). The effect of psychological intervention on perceived pain in children undergoing venipuncture. *Biomedical Papers of the Medical Faculty of the University Palacký, Olomouc, Czechoslovakia*, *155*(2), 149–154.

Taddio, A., Appleton, M., Bortolussi, R., Chambers, C., Dubey, V., Halperin, S., ... Shah, V. (2010). Reducing the pain of childhood vaccination: An evidence-based clinical practice guideline. *Canadian Medical Association Journal*, *182*(18), E843–E855.

Truba, N. & Hoyle, J.D. (2014). Pediatric pain. *Journal of Pain Management*, *7*(3), 235–248.

Tüfekci, F.G., Celebioğlu, A. & Küçükoğlu, S. (2009). Turkish children loved distraction: Using kaleidoscope to reduce perceived pain during venipuncture. *Journal of Clinical Nursing*, *18*(15), 2180–2186.

Wente, S.J. (2013). Nonpharmacologic pediatric pain management in emergency departments: A systematic review of the literature. *Journal of Emergency Nursing*, *39*(2), 140–150.

Wong, C., Lau, E., Palozzi, L. & Campbell, F. (2012). Part 1—Pain assessment tools and a brief review of nonpharmacological and pharmacological treatment options. *Canadian Pharmaceutical Journal*, *145*(5), 222–225.

*Strengthening Research Capacity and Disseminating New Findings in Nursing and Public Health – Malini et al. (Eds)*
*© 2018 Taylor & Francis Group, London, ISBN 978-1-138-50066-2*

# Factors influencing nutritional status of children aged under five based on weight-for-height using the CART method

H. Yozza & I. Rahmi
*Faculty of Mathematics and Natural Science, Andalas University, Padang, Indonesia*

H.A. Rahmy
*Faculty of Public Health, Andalas University, Padang, Indonesia*

ABSTRACT: This study is intended to determine factors that affect the nutritional status of children under five years of age in Padang City, West Sumatra, based on weight-for-height using a Classification And Regression Tree (CART). The study was carried out in four districts of Padang City. A total of 311 children under five years of age were examined. The children's nutritional status was assessed using weight-for-height $z$-scores in compliance with the World Health Organization standard for child growth. Those factors that influence children's nutritional status were hypothesised as being gender, age, family income, maternal education level, number of children, and score of maternal knowledge about nutrition. Data was analysed using a tree-classification method, namely, the CART method. It was found that three variables affected children's nutritional status: age, family income, and maternal knowledge about nutrition.

*Keywords*: nutritional status, children under five years, CART, weight-for-height

## 1 INTRODUCTION

Nutritional status is defined as being the body condition influenced by diet, levels of nutrient, and the ability of those levels to maintain normal metabolism. For children, this can be assessed from weight-for-age, height-for-age, weight-for-height, or other indicators that are compared with the standard data for adequate nourishment. Children's nutritional status is one indicator for assessing the entire population's health status. Thus, monitoring children's nutritional status is a fundamental instrument for measuring the health of a population (Monteiro et al., 2009). Nutritional status problems in children affect their future life and may lead to morbidity and death. Children's nutritional status is one predictor for their survival. Several studies (e.g. Schroeder & Brown (1994)) conducted on the relationship between nutritional status and child survival found that there is a strong relationship between nutritional status and child mortality.

Health research that was conducted in 2013 by the Department of Health in Padang, the capital city of West Sumatra Province in Indonesia, showed that the prevalence of malnutrition in Padang was high. This research found that Padang had recorded 11.7% prevalence of underweight (weight-for-age), 33.7% prevalence of stunting (height-for-age), and 9.1% prevalence of wasting (weight-for-height). Considering that this issue is related to child morbidity and mortality, this situation needs improving. Factors associated with the nutritional status of children need to be identified as points of departure to improve the situation. Therefore, the aim of this study is to determine factors that influence the nutritional status of children under five years of age in Padang City, West Sumatra.

## 2 METHOD

### 2.1 *Study Design*

This is cross-sectional research that intends to identify the influencing factors on the nutritional status of children aged under five. To assess children's nutritional status, the weight-for-height *z*-score was determined according to the World Health Organization standard for child growth and used as an indicator. Based on this indicator, nutritional status was classified into four categories, namely, 'severely wasted', 'wasted', 'normal', and 'overweight'. Those factors impacting on children's nutritional status were hypothesised as being age, gender, family income, mother's education level, number of children, and score of mother's knowledge about nutrition.

### 2.2 *Sample/participant/sampling technique*

The population of interest in this study was children under five years of age in Padang City, West Sumatra, Indonesia. A survey was conducted at several *Posyandu* (maternal and child health services), day care sessions, and housing in four districts in Padang. A total of 311 under-five children participated in this survey. In order to identify the socio-demographic and economic characteristics of the children and their families, a parent (mother) was also involved as a respondent. Participants were selected by using the purposive sampling technique.

### 2.3 *Data collection and analysis*

This study uses the primary data collected from July to September 2017. The children's weight and height were measured directly by the surveyors. The children's nutritional status was assessed by calculating the weight-for-height *z*-score using the following formula:

$$z-Score=\frac{(actual\ weight-reference\ median)}{reference\ standard\ deviation}$$

There are four categories of nutritional status: severely wasted, wasted, normal, and overweight. The cut-off points for these categories are: $z$-score $< -3$ SD for severely wasted; $-3$ SD $<$ $z$-score $< -2$ SD for wasted; $z$-score $> 2$ SD for overweight. Socio-economic information (gender, age, family income, number of children, mother's education level, mother's job status) were collected through personal interview techniques by means of a questionnaire as a guideline. The mother's knowledge about nutrition was obtained by adding the scores from 15 questions about nutrition.

There have been many studies on the nutritional status of children under five years old but most of them use bivariate analysis technique. This current research uses a modelling technique to show the application and development of a model of nutritional status of children under five years of age. The method used is a tree-classification technique, which is more informative due to its comprehensive analysis of all variables involved in the model formation.

In this research, factors associated with the nutritional status were identified by means of a Classification And Regression Tree (CART), which is a decision-tree algorithm that can be used for classifying or regressing predictive modelling problems. This is a non-parametric statistical method that models the relationship between a dependent variable (response) and other predictor variables that were assumed to influence it. The model is obtained by recursively partitioning the dataspace (called the parent node) into two sub-dataspaces (called child nodes) based on predictors involved in the analysis. This splitting procedure will be continued until the stopping rule is fulfilled. The terminal node is a node that needs no further splitting. The predictive response is obtained within each terminal node. As a result, the partitioning procedure can be represented graphically as a decision-tree diagram so that it is easily interpreted. Goodness-of-fit of the resulting model is assessed by calculating the percentage of observations that are correctly classified by the model. Data exploration and data analysis are conducted using SPSS statistical software.

### 2.4 *Ethical consideration*

This study has been reviewed and approved by Research Institution and Community Service of Andalas University.

## 3 RESULTS

### 3.1 *Data exploration*

Data consists of 311 observations. Table 1 shows the distribution of children into each nutritional status and indicates that most of the children are in the normal (45%) to overweight (23%) status range. However, 14.5% of them are categorised as severely wasted.

Socio-demographic and economic characteristics of the children and their families can be described as follows. Females account for 54% of the children. The highest education level for 55% of the mothers is senior high school, while only 24% of the mothers work. It is also shown that 44% of the children come from a low-income family (less than two million rupiah per month) while 46% are from families with an income of 2–5 million rupiah per month.

A chi-squared test was used to predict the association between each categorical predictor (gender, maternal education level, maternal job status, and family income) and the nutritional status. The results are shown in Table 2.

It can be seen that there are two predictors significantly associated with nutritional status (i.e. maternal job status and family income), while the two other predictors are independent of the nutritional status.

### 3.2 *CART result*

A tree diagram is constructed as the result of the CART method, as shown in Figure 1. This diagram illustrates that the root node (node 0) consists of 311 observations. Three splitting variables are shown in this diagram: age, family income, and maternal knowledge about

Table 1. Child distribution according to nutritional status.

| Nutritional status | Frequency | Percentage |
|---|---|---|
| Severely wasted | 45 | 14.5 |
| Wasted | 54 | 17.4 |
| Normal | 140 | 45.0 |
| Overweight | 72 | 23.2 |
| Total | 311 | 100.0 |

Table 2. Results of chi-squared test for independency.

| Variable | $\chi^2$ statistic | p-value | Conclusion |
|---|---|---|---|
| Gender vs nutritional status | 0.335 | 0.953 | Independent |
| Maternal education level vs nutritional status | 9.773 | 0.369 | Independent |
| Maternal job status vs nutritional status | 6.693 | 0.094 | Dependent |
| Family income vs nutritional status | 11.757 | 0.072 | Dependent |

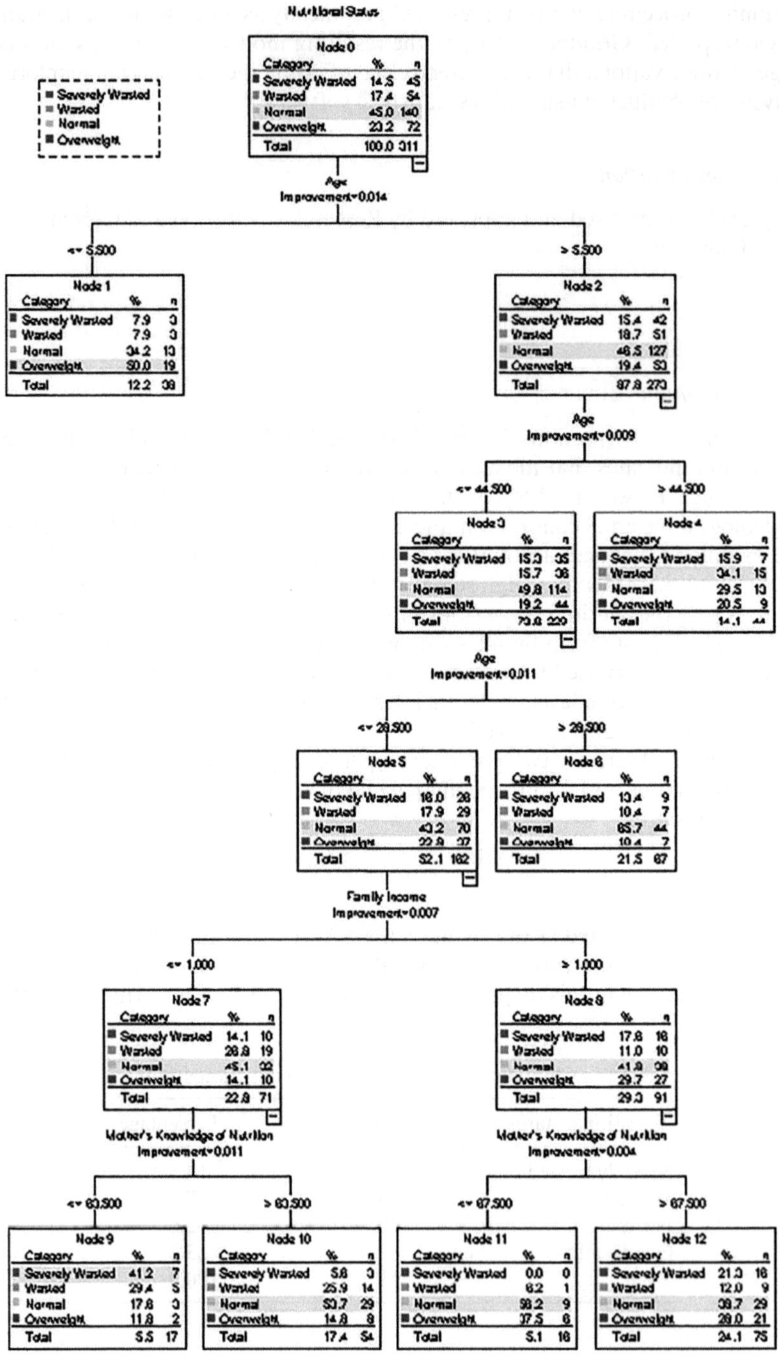


Figure 1. Tree diagram illustrating CART method.

nutrition. It means that these three variables can be considered as factors in this research that impact on the nutritional status of children under five years old.

Furthermore, it is also shown that the CART algorithm gave six terminal nodes, so that the under-five children can be put into the following groups: children under 5.5 months old, who are predicted to have overweight nutritional status; children over 44.5 months old, who tend to be of wasted nutritional status; children 28.5–44.5 months old, predicted to have normal nutritional status; children 5.5–28.5 months old and who come from families with income more than 2 million rupiah per month, who also tend to have normal nutritional status; children 5.5–28.5 months old who come from families with incomes of less than 2 million rupiah per month with poor maternal knowledge about nutrition, who have a greater probability of being categorised as being severely wasted children.

Children 5.5–28.5 months old and who come from families with incomes of less than 2 million rupiah per month, and whose mothers have a good knowledge about nutrition, tend to be of normal nutritional status. In this model, 75.1% of observations are classified correctly.

## 4 DISCUSSION

The results previously illustrated show that maternal knowledge about nutrition is one of the important factors that influence the nutritional status of children under five years old in Padang. The prevalence of severe wasting occurred in children who come from families with low incomes and who have mothers with poor knowledge about nutrition. However, mothers with sufficient knowledge about nutrition tend to have children with normal nutrition levels, even if the family income is low. To address this disparity, the mothers' knowledge about proper nutrition for their children has to be improved. One improvement initiative that the city government has pursued is through the *Posyandu* programmes.

*Posyandu*, which stands for *Pos Pelayanan Terpadu* [Integrated Service Post] (for maternal and child health), is an Indonesian government programme that provides an integrated community-based health service. This service provides for the community by involving the community in its implementation. *Posyandu* provides health services to children under five years old and to pregnant women, and also provides family planning. *Posyandu* is conducted by community members under the guidance of health workers.

In practice, the implementation of the *Posyandu* programme in Padang has been running well. In addition to weighing and examining the health of children and infants, the community (in this case the mothers) is given education about nutrition and family health. However, surveys conducted during this study identified that optimal participation from targeted community members was not in evidence. Therefore, greater efforts are needed to involve targeted community members in *Posyandu*.

One possible solution is to provide supplementary food (snacks) or gifts to children brought to *Posyandu*. Of course, the city government needs to allocate more funds through the health department. Supporting funds from surrounding companies (through their Corporate Social Responsibility programmes) can also be pursued for this purpose. In addition, *Posyandu* could hold other social activities, such as competitions, to attract the targeted community to pay more regular visits.

## REFERENCES


Breiman, L., Friedman, J.H., Olshen, R.A., & Stone, C.J. (1993) Classification and Regression Tree. New York: Chapman & Hall.

Kabeta, A., Belagavi, D. & Gizachew, Y. (2017). Factors associated with nutritional status of under-five children in Yirgalem Town, Southern Ethiopia. *IOSR Journal of Nursing and Health Science*, *6*(2), 78–84.

Kim, H. & Loh, W.Y. (2001). Classification trees with unbiased multiway splits. *Journal of the American Statistical Association*, *96*(454), 589–604.

Mazarina, D. (2010). Analysis of influential factors on nutritional status of under-fives in rural areas. *Journal of Technology and Vocational*, *33*(2), 183–192.

Monteiro, C.A., Benicio, M.H.D., Konno, S.C., Silva, A.C.F., Lima, A.L.L. & Conde, W.L. (2009). Causes for the decline in child under-nutrition in Brazil, 1996–2007. *Rev Saude Publica, 43*(1), 35–43.

Rahmi, I.H.G. & Yozza, H. (2013). Implementation of the QUEST method to identify accreditation assessment components differentiating school accreditation (case study of SMA/MA in West Sumatra). *Journal of Mathematics Journal of Theory and Applied Mathematics*, *12*(1). 29–40.

Schroeder, D.G. & Brown, K.H. (1994). Nutritional status as a predictor of child survival: Summarizing the association and quantifying its global impact. *Bulletin of the World Health Organization*, *72*(4), 560–579.
World Health Organization (WHO). (2011). *World health statistics.* Geneva, Switzerland: World Health Organization.

*Strengthening Research Capacity and Disseminating New Findings in Nursing and Public Health – Malini et al. (Eds)*
*© 2018 Taylor & Francis Group, London, ISBN 978-1-138-50066-2*

# Factors related to bullying behavior among school-age students in Padang, West Sumatra

Hermalinda, Deswita & E. Oktarina
*Faculty of Nursing, Andalas University, Padang, West Sumatra, Indonesia*

ABSTRACT: This research is aimed to analyze factors that are related to bullying behavior in Junior High School students in Padang city. This is a descriptive analytic research using a cross sectional study approach. Samples were 340 students that were taken through proportional random sampling technique. Data were collected by using a questionnaire eliciting students' charateristics, school environment, and family factors. Adolescent Peer Relation Instrument (APRI) was applied to measure bullying behavior. The data were analyzed by using Mann Whitney and Kruskall Wallis Test. The analyses showed that there was a significant relationship between parenting style and bullying behavior. School environmental factors such as the school's atmosphere, group support, and teacher assistance were insignificantly related to bullying behavior. The school pressure had not related to bullying behavior. Social and institutional interventions for preventing bullying behavior among teenagers are expected to be presented at schools which involve social groups, teachers, and parents.

## 1 INTRODUCTION

In Indonesia, bullying cases become the top issue the community complained at most. From 2011 to August 2014, the Indonesian Commission on Child Protection (KPAI) recorded 369 complaints related to the bullying issues. The bullying cases formed 25% of the total complaints from 1,480 cases in the field of education. According to the International Center for Research on Women research report in 2014, the prevalence of violence concerning children at schools was very high where 84% students had experienced violence at school and 75% said that it happened in the last six months.

From 1,075 school age children in the city of Pelotas, RS Brazil, there were 17.6% children with bullying behavior. The most common type of bullying is verbal, followed by physical, emotional, and sexual harassments (Moura, Catarina, Cruz, & Quevedo, 2011). A study that was conducted on 1,229 children in seven cities in USA revealed that nearly a half of the respondents were against the bullying; one fourth discussed it with adults; 20% did not do anything; and only 8% spoke with the bullies (Brown, Birch, & Karcherla, 2005).

Children who become the perpetrators or even the victims of bullying suffered from such negative impacts as depression, anxiety, substance abuse, low social functions, poor academic achievement, and getting less of attention. Repeated bullies and victims are at risk for suicide (Tsitsika et al., 2014). Victims directly reported physical problems such as sore throat, colds, and coughs. Children also experienced such psychosomatic problems as decreased appetite and fear to go to school (Wolke & Skew, 2012). young female victims indicate specific problems such as headaches and sleep disorders (Biebl, Dilalla, Davis, Lynch, & Shinn, 2011). Ultimately, bullying can result in depression (Due, Damsgaard, Lund, & Holstein, 2009).

Bullying is a complex and noticeable construction that can affect the quality of life. It occurs due to a number of factors that children experience such as psychological, cognitive, and emotional factors as well as the influence of specific surroundings such as parents' expectations and socioeconomic status (Liu & Graves, 2011). A longitudinal study of Individual Adolescent's Tracking Survey (TRAILS) in the Netherlands obtained the result that preschool behavior, emotional and movement problems, socioeconomic status, and family divorce were associated with bullying at later age (Jansen, Veenstra, Ormel, Verhulst, & Reijneveld, 2011).

## 2 METHOD

### 2.1 *Study design*

This is a descriptive analytic study with a cross sectional study approach.

### 2.2 *Sample and sampling technique*

The population of this research is all students of four junior high schools in Padang city. Samples were taken by using proportional random sampling technique. As the result, the final sample size was 340 students.

### 2.3 *Ethical considerations*

This research has obtained ethical consideration from the Ethics Committee of Faculty of Medicine Andalas University. Before collecting the data, reseachers gave an explanation to the respondents about research prosedure in which they were then asked to participate and also guaranteed the secrecy of the respondents' identity.

### 2.4 *Data collection and analysis*

Data were collected by using questionnaires that can be filled in by teenagers in the researchers' presence. The questionnaire consists of multiple components extracting the following information: Family Factors by using Parent's Perceptions Questionnaire (PPQ), which consists of 20 statements. Among them, there are ten statements about responsive and demanding parenting. There are four answer choices: 4, always; 3, often; 2, sometimes; and 1, never. (Pasquali, Gouveia, & Andrade, 2012); Environmental factors including school atmosphere that are scaled in six statements, pressure in the school in four statements, group support in three statements and, teachers' support in four statements. Answer choices are ranging from strongly disagree;1 to strongly agree; 5. The Adolescent Peer Relations questionnaire Instrument (Parada, 2000) consists of two sections: section A investigating the behavior of bullying and section B concerning the victims of bullying. Each section consists of 18 statements with alternate answers: never happened; 1, sometimes; 2, once or twice in 1 month; 3, once a week; 4, more than once a week; 5, and every day; 6.

Mann Whitney and Kruskall Wallis tests were employed to examine the relationship between the parenting style and school environment factors. Findings were considered statistically significant if the p value was $\leq 0.05$.

## 3 RESULT

### 3.1 *Analysis of the relationship between the characteristics of adolescents/families and bullying behaviour*

Table 1 indicates that students raised by authoritarian parents show the highest mean values among all styles. The statistical test result shows that there is a significant difference between parenting style and bullying behavior.

In addition, Table 2 points out a significant relationship between school environment and bullying behavior. There is a week correlation between atmosphere of the school (0.30), teachers support (0.20), and group support (0.26) and bullying behavior. However, there is no relationship between school pressure and bullying behavior.

Table 1. Relationship between family factors and bullying behavior.

| Variable | N | The bully median (min-max) | *p* value |
|---|---|---|---|
| Parenting style | | | |
| Authoritarian | 53 | 27.00 (18–66) | 0.001 |
| Authoritative | 133 | 23.00 (18–50) | |
| Uninvolved | 107 | 25.00 (18–57) | |
| Permissive | 47 | 26.00 (18–108) | |

Table 2. Relationship between school environment and bullying behaviour.

| | | The bully | | The victims | |
|---|---|---|---|---|---|
| No | Variable | *r* | *p* value | R | *p* value |
| 1. | School Environment | | | | |
| | School Atmosphere | 0.306 | 0.000 | 0.266 | 0.000 |
| | School Pressure | 0.037 | 0.492 | 0.040 | 0.460 |
| | Teacher Support | 0.202 | 0.000 | 0.322 | 0.000 |
| | Group Support | 0.264 | 0.000 | 0.288 | 0.000 |

## 4 DISCUSSION

### 4.1 *Relationship between family factor and bullying behavior*

The analysis shows that bullying behavior is more common in children with authoritarian parenting than those with authoritative, uninvolved, and permissive parentings. The statistical test also confirms a significant difference in bullying behavior between authoritative parenting and all other parenting styles. As one of the family responsibilities is to develop behavior that is culturally appropriate to children, the children learn to behave in a way that is expected according to their position in the family (Hockenberry & Wilson, 2013). Family factors are then associated with behavioral problems in adolescents (Driessens, 2015). There is a relationship between family violence (such as physical injury or witnessing the violence of family members) and aggressive behavior in children (Morbidity and Mortality of Weekly Report, 2011). Affection and parental behavior are consistently the influencing factors to a child's ability to engage in healthy relationships at all stages of development. Bullying behavior can be an indicator of the risk of various mental health disorders in adolescents (Liu & Graves, 2011).

Parenting style are factors that increase the risk of bullying behavior. Children raised by authoritarian parents that employ harsh punishment, lack of freedom, and a strict environment tend to have a high potential to develop bullying behavior. Parenting styles that do not allow children to develop their decision-making or opinion-raising skills will increase the risk of children for becoming the victims (Our Kids Network, 2014).

In the family context, especially parental behavior, there is a significant difference between the bullying group, the victims, and the groups who are not involved in bullying behavior. Parents of the bullying groups and victims more often exercise negative discipline and psychological controls in their parenting and obtain a lack of acceptance from their children. The bullies have parents who lack autonomy and control over their children. Parenting styles that apply positive and permissive discipline do not have a significant relationship with children's bullying behavior (Velki, 2012).

Previous studies have shown that children from parents who express anger and feel that their children are troublesome or have difficulties in nurturing the children tend to develop bullying behavior. Children who always complete homeworks and have parents who want to meet and talk to their friends tend to have a lower possibility of becoming bullies (Shetgiri, Lin, Avila, & Flores, 2012). In Egypt, Morocco, and Tunisia, students who reported that their parents checked their homework, understood, and knew how students spent their spare time rarely became the victims of bullying groups (Abdirahman, Fleming, & Jacobsen, 2008).

Family characteristics are associated with children's involvement in bullying behavior at later ages (Jansen et al., 2011). Other studies have also identified that there is a relationship between bullying and family background/structure (Carlerby, Viitasara, Knutsson, & Ga, 2013). Children who lived with a single parent were more potential to develop bullying attitude as compared to those who lived with both parents. Parents who always spent time with their children every week reduced the risk of the children's bullying behavior while parents who frequently quarreled with their children increased the risk (Wolke, Woods, Bloomfield, Karstadt, & Lane, 2001).

### 4.2 *Relationship between school environtment and bullying behavior*

The analysis of this study found out that there is a meaningful but week relationship between school atmosphere (r: 0.25), groups support (r: 0.22), teachers' support (r: 0.24) and bullying behavior. However, there was no significant relationship between school pressures of school and bullying behavior (Shaheen et al., 2014). Moreover, school factors and involvement in extracurricular activities in schools are associated with behavioral problems in adolescents (Driessens, 2015).

In terms of performance at school, uninvolved children (i.e. neither a bully nor a victim) have better achievements than the bully and victims do. They also feel safer at school than victims and bullying perpetrators do. In addition, bullying actors have the most negative opinion and the victims have the least negative opinion at the school atmosphere. The victims of bullying will feel that school is susceptible and mostly negative but children who are not involved will feel that their school has a positive atmosphere (Velki, 2012).

During the early stages of adolescence, the pressure to have a group gets stronger. Teenagers consider that having a group is important because they feel that becoming a part of a group can give them a status. Teenagers will try to symbolize their membership to the group through appearances by wearing the same clothes, applying makeup, and styling their hair. Language, music, and dance signify the exclusive group culture on teenagers. Teenagers who attempt to show individual differences are unacceptable and most possibly exiled from the group (Hockenberry & Wilson, 2013). Relationships in groups also affect bullying behavior. Lack of relationships in positive groups and the large number of friends will lead to a sense of popularity and position of power (Our Kids Network, 2012).

A total of 359 teachers participated in a research about their perception and action to prevent bullying behavior. A majority of the teachers (86.3%) talked seriously with perpetrators and victims of bullying. Less than one-third of the teachers (31.2%) arranged a learning time in class to discuss bullying or involve pupils to fight bullying behavior. The teachers considered that every time bullying occurred, a surveillance needs to be done to the students and conditions that discourage bullying activities in the environment needs providing (Joseph, James, Susan, & Jeanne, 2003).

## 5 CONCLUSIONS

There are significant differences in bullying behavior based on parenting styles and school environtment factors but there is no significant relationship between school stress and bullying behavior. The presence of an intervention involving family, friends, and teachers which nurture positive communication, self-expression, cooperation, and connectedness is expected in order to develop a strong protective elements to resist bullying behavior. Nurses are expected to develop investigations on violent behavior and bullying in children.

## REFERENCES


Abdirahman, H., Fleming, L.C., & Jacobsen, K.H. 2008. Parental involvement and bullying among middle—school students in North Africa. *Journal Health Mediterranean Eastern, 19*(3), 227–234..

Biebl, S.J.W., Dilalla, L.F., Davis, E.K., Lynch, K.A., & Shinn, S.O. (2011). Longitudinal associations among peer victimization and physical and mental health problems. *Journal of Pediatric Psychology*, 36(8), 868–877.

Brown, S.L. Birch, D.A., & Karcherla, V. 2005. Bullying perspective attitude and recommendations of 9 to 13 year old attending health education centers in the United State. *Journal Science Health*, 75(10), 384–82.

Carlerby, H., Viitasara, E., Knutsson, A., & Ga, K.G. 2013. How Bullying involvement is associated with the distribution of parental background and with subjective health complaints among Swedish boys and girls. *Social Indicator Research*, 111(3), 775–783.

Driessens, C.M.E.F. 2015. Extracurricular activity participation moderates impact of family and school factors on adolescents ' disruptive behavioural problems. *BMC Public Health*, 15, 1–14.

Due, P., Damsgaard, M.T., Lund, R., & Holstein, B.E. 2009. Is bullying equally harmful for rich and poor children ? A study of bullying and depression from age 15 to 27?. *Journal of Public Health*, 9(5), 464–469. doi.org/10.1093/eurpub/ckp099.

Hockenberry, M. J & Wilson, D. 2009. *Wong's essential of pediatric nursing*, 8th Edition. St. Louis: Mosby Year Book.

Jansen, D.E.M.C., Veenstra, R., Ormel, J., Verhulst, F.C., & Reijneveld, S.A. 2011. Early risk factors or being a bully, victim, or bully/victim in late elementary and early secondary education. The longitudinal TRAILS study. *BMC Public Health*, 11(1), 440.

Joseph, A., James, H., Susan, K., & Jeanne, B. 2003. Teacher Perceptions and Practices Regarding School Bullying Prevention. *The Journal of school Health*, 73(9), 347–354.

Liu, J., & Graves, N. 2011. Childhood bullying: A review of construct, context and nursing implication. *Public Health Nurse*, 28(6), 556–568.

Morbidity and Mortality Weekly Report (MMWR). 2011. *Bullying among middle school and high school students—Massachusetts, 2009*. Morbidity and Mortality Weekly Report (MMWR), 60(15), 465–471.

Moura, D.R. De, Catarina, A., Cruz, N., & Quevedo, L.D.Á. 2011. Prevalence and characteristics of school age bullying victims. *Journal de Pediatria*, 87(1), 19–23.

Our Kids Network (OKN). 2012. Bullying prevention and support among school age children and youth: *A review of the literature*. Halton Kids Our Kids Network, School Year Committee.

Pasquali, L., Gouveia, V.V., & Andrade, J.M. De. 2012. Perceptions of parents questionnaire: Evidence for a measure of parenting styles. Paidéia, 22(52), 155–164.

Parada, R.H. 2000. *Adolescent Peer Relations Instrument: A theoretical and empirical basis for the measurement of participant roles in bullying and victimization of adolescence: An interim test manual and a research monograph: A test manual.* Penrith South, DC, Australia: Publication Unit, Self-concept Enhancement and Learning Facilitation (SELF) Research Centre, University of Western Sydney.

Shetgiri, R., Lin, H., Avila, R.M., & Flores, G. 2012. Parental characteristics associated with bullying perpetration in US children aged 10 to 17 years. *American Journal of Public Health*, 102(12), 2280–2287.

Tsitsika, A.K., Barlou, E., Andrie, E., Dimitropoulou, C., Tzavela, E.C., Janikian, M., ... Tsolia, M.2014. Bullying behaviors in children and adolescents: "an ongoing story," *Frontiers in Public Health*, 2, 7.

Velki, T. 2012. A Comparison of individual characteristics and the mupltiple contexts for children with different bullying status: An ecological perspective. *International Journal of Arts & Sciences*, 5(7), 89–112.

Wolke, D., Wolke, D., & Skew, A.J. 2012. Family factors, bullying victimisation and wellbeing in adolescents. *Longitudinal and Life Course Studies*, 3(1), 101–119.

Wolke, D., Woods, S., Bloomfield, L., Karstadt, L., & Lane, C. 2001. Bullying involvement in primary school and common health problems. *Archives of Disease in Childhood, 85*(3), 197–201.

*Strengthening Research Capacity and Disseminating New Findings in Nursing and Public Health – Malini et al. (Eds)*
*© 2018 Taylor & Francis Group, London, ISBN 978-1-138-50066-2*

# Family-centred care needs of mucositis management in children with cancer

I. Nurhidayah, H.S. Mediani, S. Hendrawati & A. Mardhiyah
*Department of Pediatric Nursing, Faculty of Nursing, Universitas Padjadjaran, Indonesia*

ABSTRACT: Mucositis is a side effect of chemotherapy that occurs in children with cancer. Participation of parents in Family-Centred Care (FCC) is an important part of the management of mucositis. However, in Indonesia, a study that examines the FCC of mucositis management in children with cancer has not yet been established. Thus, the aim of this study was to analyse the needs of FCC in the management of chemotherapy-induced mucositis in children with cancer. This study was qualitative and conducted in one hospital in Bandung. Ten families with children with cancer participated. Data was collected through focus group discussions and analysed by using theme analysis. There were five themed needs of parents: information and education needs; emotional needs; psychosocial needs; cooperation and collaboration needs; physical needs. Nurses and other healthcare providers should be familiar with the parental needs in chemotherapy-induced mucositis care, particularly in relation to information and emotional needs.

*Keywords*: Children with cancer, family centered care need, mucositis management

## 1 INTRODUCTION

Cancer is a collection of abnormal cells that grow continuously, unlimited and uncoordinated with the surrounding tissue and not functioning physiologically (Price & Wilson, 2005). Cancer cells can be mutagenic cells because of genetic mutations in germ cells as well as in somatic cells. Various factors have a role to play, including heredity and environmental factors (Baggott et al., 2002). Mutagenic cells have infiltration properties (infiltrating the surrounding tissue) as well as destructive properties (damaging the surrounding tissue). The cells divide uncontrollably and eventually attack other cells. This will lead to a series of changes in cell metabolism that will ultimately interfere with the physiological functions of the body (Price & Wilson, 2005).

Strategies are needed to minimise the effects of chemotherapy, particularly mucositis among others (Ropi et al., 2015). The family role is vital, especially that of parents supporting the treatment of their children with cancer. The role of parents is very important and determines the success of treatment and care as well as the cancer survival rate in children. According to Hockenberry and Wilson (2011), caring for children can be optimised by involving the family. Children should not be separated from the family, and one of the functions of the family is to provide healthcare for their children.

When children have health problems, especially chronically ill children with, for example, cancer, families are required to provide the best care for their children. Under these conditions, parents attempt to provide optimal care, especially when children are experiencing the effects of chemotherapy, such as mucositis, during hospitalisation. Nurses, as caregivers, should be aware of the importance of family engagement in working actively with other healthcare workers. In the nursing care of the child, the family have a central and specialised role to take care of the child's physical needs, to educate the child, and to be responsible for the psychological and emotional well-being of the child.

Nurses have an important role in empowering family. Hockenberry and Wilson (2011) stated that family-centred care is a concept for empowering and enabling. Enabling means creating opportunities and ways for all family members to be involved, while empowering means FCC creates interaction between nurses and families in such a way that the family maintains control of, or gets the feeling of being able to control, their lives and all aspects of the positive changes of the impact of aid behaviour.

## 2 METHODS

The design of this study was qualitative. The study was conducted in one hospital in Bandung. There were ten families with children with cancer who participated in this study. Data was collected through Focus Group Discussions (FGD). Inclusion criteria for participants were parents who had children with cancer who were undergoing treatment. The data was analysed by theme analysis from Miles and Huberman (2014), which divides the steps in the data analysis activities into data collection, data reduction, presentation of data (display data), and conclusion or verification.

## 3 RESULTS

There were several parental needs observed in the management of mucositis in child cancer, namely, emotional needs, physical needs, information and education needs, psychosocial needs, and the need for collaboration and cooperation.

a. Emotional needs
Emotional need is a parent's need to gain belonging, comfort in times of stress, and understanding of the child's condition. Parents had confusion and anxiety about the condition of their children who have mucositis.

> "... the child experience drop in health conditions nausea performed, vomiting, fall out, mouth sores, red gums until blisters ... sometimes confused as to why ... and worried getting worse..." (P2).

b. Physical needs
Physical need was the parent's need to see a child with cancer feeling physically free from pain, experiencing optimal growth, and having the ability to perform daily activities. This need is important as a basic understanding of parents in managing the mucositis experienced by children.

c. Information and education needs
Parents need information about the management of mucositis in children with cancer to reduce their confusion and anxiety about the condition of the child. General information about the management of mucositis in children with cancer is the information most sought-after by parents. They also need written information and education in managing mucositis in children with cancer. Parents were trying to find information from nurses and doctors in dealing with the mucositis complaints of their child, as indicated by the following quotes:

> "... like to be given information by the nurse to overcome the complaint... we were first asked, just told by nurse or doctor... if not asked will not be told..." (P1).
>
> "... must ask first... need information on how to overcome child mouth sores... so parents know handling..." (P2).
>
> "... yes, I have to ask first... if not, will not be told... because if not asked thought it's okay, no complaints... if we asked it will be noticed..." (P5).
>
> "... from a doctor or nurse, there is nothing to give detailed information..." (P8).

d. Psychosocial needs
Psychosocial needs were the needs of parents associated with feelings in relation to self-assessment, competence, and feelings of respect. This need is related to family

relationships, and acceptance within the community. The most important psychosocial need is social support from families, nurses or doctors who care for their children, as well as from fellow parents who have children with cancer, as indicated by the following parent disclosures:

"... I and my husband as parents of course need support from family... besides that, I am so happy if a nurse tells us how to treat child mouth sores, but we must actively ask..." (P7).

"... often chatting with same parents whose children have the same problem as my children..." (P7).

e. Collaboration and cooperation needs

Parents want good collaboration between nurses and doctors in caring for their children. In addition, parents also want to be involved or empowered in providing care for their children. Here are some parents' quotes:

"... yes, the cooperation of nurses and doctors was very important; doctors know the state of their patients from nurses..." (P1).

"... nurses and doctors should cooperate and be alert... I am happy if nurse can involve parents when they give nursing care to my children..." (P10).

## 4 DISCUSSION

Family-centred care is needed in mucositis management to improve the quality of life in children with cancer. The most family-centred care need was information and education. Emotional needs were the second requirement that parents had. This study was in line with research conducted by Kerr et al. (2007): a review of literature published between 1993 and 2001 concerning the supportive care needs of parents of children with cancer, which showed that information is the most supportive care need. The results of the study by Chen et al. (2014) on 102 caregivers of oral cancer patients in Taiwan also demonstrated that information is the most difficult supportive care requirement.

The differences in the needs of parents with children with cancer can occur because the needs are dynamic throughout the spectrum of cancer experiences, ranging from pre-diagnosis to loss (Kerr et al., 2007). In addition, the differences are also influenced by demographic, social and individual factors that affect how a person responds to a cancer diagnosis and its impact on his or her life American Cancer Society (2016).

The most difficult information and education need experienced by parents was the need to obtain information about the oral and dental care that can be provided to help manage mucositis in children with cancer. This was followed by the parents' need for written information about mucositis and its effects on children with cancer. Parents also need written information, such as leaflets, posters, and other media, about the management of mucositis in children with cancer as a provision for parents to care for their children.

This study also indicated that parents are trying to find various ways to overcome post-chemotherapy complaints. Mucositis can cause chapped and bleeding lips, dry saliva, sore throat, and even bleeding gums. It was revealed that the family want service providers, especially nurses, to manage the side effects of chemotherapy, especially mucositis, to improve the quality of life in children. Parents expected to get information and education for the management of mucositis in their children because parents reported that having a lack of information and education related to how they managed mucositis in children with cancer. Information was notified or explained if parents asked first, but was not explained in detail.

Information and education can be a strength for parents in terms of empowerment and parental involvement in caring for their children. Parental empowerment in FCC facilitates more effective information transfer, both information needed by healthcare workers in making care plans, as well as that for families about the care needed by the patient. Thereby, families become more confident, feel appreciated, and feel that their opinion is taken into account in the implementation of care. Children treated with family involvement also become calmer, and less anxious, and the healing process becomes faster. Families are also educated to be self-reliant and understand the principles of healthcare so that they can take preventive and home care measures.

Parents also need emotional support. Almost half the parents feel fear when the child's mucositis worsens. Parents were afraid if the child's condition was getting worse. Mucositis causes a loss of appetite and the child's weight decreases; the child can become moody and may not sleep adequately. Basically, parents have the most important role in the process of adjusting to the physical changes that occur due to cancer (Wenar & Kerig, 2008, cited in Puspita & Ludiro, 2013). This change can increase parents' anxiety and vulnerability. Parents feel loss of control and confused. Parents need new information, new skills, or to seek help to meet their needs (Fitch, 2008).

The need to cope with pain (e.g. headache, face, around the mouth, pain swallowing, and difficulty chewing) that the child feels as a result of the mucositis they experience is necessary. Parents need information and support for these physical needs. Children can overcome their physical problems. This situation can increase anxiety in the parent. It requires cooperation with nurses in providing care for the children. Physical needs include a parent's need to see a child with cancer feeling physically free of pain, getting optimal nutrition, and having the ability to perform daily activities. This need is as important as a basic understanding for parents in managing the mucositis experienced by children (Kerr et al., 2007).

In caring for the child in the hospital, health professionals need to involve the parents, to monitor the symptoms, the testing and the treatment or therapy (Clarke & Fletcher, 2003). Parents play an enormous role in the care of their children, and nurses and other health professionals should know about this and apply the principle of FCC in the care of children with cancer. The paediatric nurse should advocate the participation of family members in order to follow up on the management of mucositis. FCC has an important role in improving the quality of life of children with cancer.

The care of the child with cancer is continuous care. Treatment of children with cancer does not stop after the cycle of chemotherapy is complete. The care of the child with cancer continues after the child is declared free from cancer; there are long-term side effects that can occur in children with cancer, such as recurrence (or cancer that reappears), the risk of death, the presence of a second primary neoplasm, neurocognitive defects, cardiovascular disease, and other organ disorders and psychosocial disorders (Oeffinger et al., 2006). Thus, parents need adequate information support so that they are encouraged to empower the child's self-management as they mature, to follow up their health conditions and reduce the risk of long-term side effects of the cancer care during childhood (Pritchard-Jones et al., 2013).

## 5 CONCLUSIONS

The role of the parents is very important and determines the success of treatment and care as well as increasing the cancer survival rate in children. From this study it was found that parents need support from health professionals, especially nurses, who have more contact time with children and their families than other health workers. Parents need information about how to deal with mucositis (related to oral care, feeding, food selection, and other information). Parents also need emotional support from nurses to be able to provide child care, have respect for each other, and collaborate between parents and caregivers. Parents wish that nurses can always provide support to families, so that families can provide optimum care for their children with cancer.

## REFERENCES

American Cancer Society. (2016). *How are childhood cancers found?*. Available at American Cancer Society Web Site: www.cancer.org.

Baggott, C.R., Kelly, K.P., Fochtman, D. & Foley, G. (2002). *Nursing care of children and adolescents with cancer* (3rd ed.). Pennsylvania, PA: Saunders.

Cancer Care Nova Scotia. (2013). *Best practice guidelines for the management of oral complications from cancer therapy*. Halifax, Canada: Nova Scotia Government.

Chang, A.M., Molassiotis, A., Chan, C.W.H. & Lee, I.Y.M. (2007). Nursing management of oral mucositis in cancer patients. *Hong Kong Medical Journal, 13*(1), 20–26.
Chen, S.C., Lai, Y.H., Liao, C.T., Lin, C.Y., Fan, K.H. & Chang, J.T. (2014). Unmet supportive care needs and characteristics of family caregivers of patients with oral cancer after surgery. *Psycho-Oncology, 23*(5), 569–577.
Cheng, K.K., Lee, V., Li, C.H., Yuen, H.L. & Epstein, J.B. (2012). Oral mucositis in pediatric and adolescent patients undergoing chemotherapy: The impact of symptoms on quality of life. *Support Care Cancer, 20*(10), 2335–2342.
Clarke, J.N. & Fletcher, P. (2003). Communication issues faced by parents who have a child diagnosed with cancer. *Journal of Pediatric Oncology Nursing, 20*(4), 175–191.
Creswell, J.W. (2009). *Research design: Qualitative, quantitative, and mixed methods approaches* (3rd ed.). Thousand Oaks, CA: Sage Publications.
Denboba, D., McPherson, M.G., Kenney, M.K., Strickland, B. & Newacheck, P.W. (2006). Achieving family and provider partnerships for children with special health care needs. *Pediatrics, 118*(4), 1607–1615.
Eilers, J. (2004). Nursing interventions and supportive care for the prevention and treatment of oral mucositis associated with cancer treatment. *Oncology Nursing Forum, 31*(4), 13–23.
Elting, L.S., Cooksley, C., Chambers, M., Cantor, S.B., Manzullo, E. & Rubenstein, E.B. (2003). The burdens of cancer therapy. *Cancer, 98*(7), 1531–1539.
Fitch, M.I. (2008). Supportive care framework. *Canada Oncology Nursing Journal.* DOI: 10.5737/1181912 × 181614.
Gatot, D. & Windiastuti, E. (2016). Treatment of childhood acute lymphoblastic leukemia in Jakarta: Result of modified Indonesian National Protocol 94. *Paediatrica Indonesiana, 46*(4), 179–184.
Hockenberry, M.J. & Wilson, D. (2011). *Wong's essentials of pediatric nursing* (8th ed.). St. Louis, MO: Mosby.
IARC. (2014). *IARC world cancer report.* Lyon, France: International Agency for Research on Cancer. Retrieved from http://publications.iarc.fr/Non-Series-Publications/World-Cancer-Reports.
Kerr, L.M.J., Harrison, M.B., Medves, J. & Tranmer, J. (2004). Supportive care needs of parents of children with cancer: Transition from diagnosis to treatment. *Oncology Nursing Forum, 31*(6), E116–E126. doi:10.1188/04.ONF.E116–E126.
Kerr, L.M.J., Harrison, M.B., Medves, J., Tranmer, J. & Fitch, M.I. (2007). Understanding the supportive care needs of parents of children with cancer: An approach to local needs assessment. *Journal of Pediatric Oncology Nursing, 24*(5), 279–293.
Miles, M.B., & Huberman, A.M. (2014). *Qualitative data analysis: A methods sourcebook* (3rd Ed.). USA: Sage Publications.
National Cancer Institute. (2016). Surveillance, epidemiology and end results program (SEER). Retrieved from http://www.seer.cancer.gov/canques/incidence.html.
Nurhidayah, I., Solehati, T., & Nuraeni, A. (2013). Mucositis score in children with cancer in Bandung. *Jurnal Keperawatan Soedirman, 8*(1). page 1–13.
Oeffinger, K.C., Mertens, A.C., Sklar, C.A., Kawashima, T., Hudson, M.M., Meadows, A.T., ... Robison, L.L. (2006). Chronic health conditions in adult survivors of childhood cancer. *New England Journal of Medicine, 355*(15), 1572–1582.
Price, S.A., & Wilson, L.M. (2005). Pathophysiology: Clinical concept of disease processes. Jakarta: EGC.
Pritchard-Jones, K., Pieters, R., Reaman, G.H., Hjorth, L., Downie, P., Calaminus, G., ... Steliarova-Foucher, E. (2013). Sustaining innovation and improvement in the treatment of childhood cancer: Lessons from high-income countries. *Lancet Oncology, 14*(3), e95–e103.
Puspita, K.S., & Ludiro, K.S. (2013). Parents' condition child with cancer patients viewed from biopsychosocial aspects based on the phase of child medicine (Descriptive study in support group as a program in The Golden Ribbon Community Dharmais Cancer Hospital Jakarta). *Journal of Faculty of Social and Political Science*. Universitas Indonesia.
Ropi, H., Mediani, H.S., Nurhidayah, I. & Mardhiyah, A. (2015). Impact of mucositis in quality of life among children with cancer in Bandung. Reports of Research Results (Not Published) Faculty of Nursing, Universitas Padjadjaran, Bandung, Indonesia.
UKCCSG-PONF. (2006). *Mouth care for children and young people with cancer: Evidence-based guidelines*. London, UK: Royal College of Paediatrics and Child Health. Retrieved from https://www.rcpch.ac.uk/sites/default/files/asset_library/Research/Clinical%20Effectiveness/Endorsed%20 guidelines/Mouth%20Care%20for%20CYP%20with%20cancer%20%28cancer%20 study%20 group/mouth_care_cyp_cancer_guidelinev2.pdf.
UKCCSG-PONF. (2006). Mouth care for children and young people with cancer: evidence based guidelines. Mouth Care Guidelines Report, Version 1, Feb 2006.

*Strengthening Research Capacity and Disseminating New Findings in Nursing and Public Health – Malini et al. (Eds)*
 *ISBN 978-1-138-50066-2*

# Analysis of phlebitis occurrence in terms of the characteristics of infusion by nurses in RSI Ibnu Sina Payakumbuh

L. Fajria, L. Merdawati & Nelviza
*Faculty of Nursing, University of Andalas, Padang, West Sumatra, Indonesia*

ABSTRACT: The purpose of this study is to determine the level of nurse compliance in using aseptic techniques to install an infusion and its relationship with the phlebitis incidence in RSI Ibnu Sina Payakumbuh. This study is a descriptive analytic study using cross-sectional methods and applying a proportional stratified random sampling technique. This research uses observation sheets based on Standard Operating Procedure (SOP) at RSI Ibnu Sina Payakumbuh as the instrument. The instrumentation of phlebitis occurrence is based on the phlebitis grade in the standard practice of the Infusion Nurses Society (INS). The results revealed a phlebitis incidence of 25.6%, while the nurse compliance was 55.8%. Bivariate test results indicate a $p$ value of 0.012 (< 0.05), which shows a relationship between nurse compliance in using aseptic techniques during the installation of an infusion and the phlebitis incidence in RSI Ibnu Sina Payakumbuh.

*Keywords*: phlebitis, infusion chracteristics, nurses

## 1 INTRODUCTION

Hospitals provide health services in the form of inpatient, outpatient and emergency services (Permenkes RI No. 340/PER/III/2010)(Depkes RI,2010). Every medical action always puts patient safety first and minimises risk. This aims to improve safety, avoid injury and improve the quality of care (Susianti, 2008). to measure continuous safety (Darmadi, 2008). Patient safety is focused on reducing nosocomial infections, decubitus and drug delivery errors, and ensuring patient satisfaction with healthcare. Nosocomial infection is an infection that occurs in patients while in hospitals or other health facilities (Darmadi, 2008). Previous nursing contact (even up to 24 hours) with the patient plays an important role in contributing to the incidence of nosocomial infection (Nursalam, 2011). A WHO study showed that approximately 8.7% of 55 hospitals in 14 countries in Europe, the Middle East, South-East Asia and the Pacific, and as many as 10.0% of hospitals in South-East Asian countries, reported cases of nosocomial infections, known as Hospital-Acquired Infection (HAI) (Putri, 2016). Among the many types of nosocomial infections, phlebitis ranks first compared to other infections (Brunner & Suddarth, 2013). Phlebitis has become an indicator of a hospital's minimum service quality, with an incidence standard of ≤ 1.5% (MOH Department of Health, 2008). In Indonesia, there is no defined incidence of phlebitis, probably because studies and publications related to phlebitis are rare. Data from Department of Health the Indonesian Ministry of Health (MOH Department of Health) in 2013 showed that the amount of phlebitis incidence in Indonesia was 50.11% for public hospitals and 32.70% for private hospitals (Department of Health-MOH, 2013). Many factors cause the occurrence of phlebitis. One factor is the aseptic or sterile techniques used during infusion. Disinfection of the area around the puncturepiercing with 70% alcohol, and sterilisation of the tools used play an important role in avoiding inflammatory complications. This can be done by, for example, washing hands before taking action and disinfecting the area around the puncture (Brunner & Suddarth, 2013). The presence of bacterial phlebitis can be a serious problem, as it predisposes to systemic complications (septicaemia). Factors contributing to the incidence of bacterial phlebitis include inadequate handwashing techniques, incomplete aseptic techniques at the time of piercing, poor catheter insertion techniques, and prolonged installation. The principle of the installation of intravenous (IV) therapy takes into account the principle of sterilisation. This is done to prevent contamination caused by intravenous needles (Rizky, 2014).

The results of research conducted by Mada et al. (2012) at Christian Hospital Lende Moripa found an inadequate application of the sterile infusion principle of 64.3% (36 persons). The application of the sterile principle involves its application prior to the installation of the IV, while performing the action, and when cleaning the appliance. Such applications are said to be sufficient if they fit the correct sterile infusion technique. Based on data from the Indonesia Infection Prevention Supervisory Team (IPCN) and the nosocomial infection control team in hospitals in 2015, an average of 28 cases, or about 5.9% of cases, of phlebitis occur in RSI Ibnu Sina Payakumbuh every month. From observations of six nurses (two nurses were assigned at random), it was found that two nurses had washed their hands both before and after intravenous treatment and four other nurses simply washed their hands after intravenous treatment. information head of the internal medicine room says that phlebitis is a complication of infusion. The role of nurses in reducing the incidence of phlebitis is very important, because the nurse is the operator who performs the infusion installation. Among the many factors that cause phlebitis, the aseptic technique performed by the nurse at the beginning of the IV should be of concern.

## 2 METHOD

This study uses descriptive analytical research with a cross-sectional approach. The aim is to analyse the incidence of phlebitis in terms of the infusion characteristics used by nurses at RSI Ibnu Sina Payakumbuh in 2016. The population is all of the nurses in the emergency unit, internal medical unit, operating room, and VIP room. A sample of 43 nurses met the criteria. The sampling used was a proportional stratified random sampling method (Arikunto, 2014). Research was for one month from 20 November to 20 December 2016. The research instrument is an observation sheet based on Standard Operating Procedure (SOP) at RSI Ibnu Sina Payakumbuh. The instrumentation of the occurrence of phlebitis is based on class phlebitis from the standard practice of the Infusion Nurses Society (INS). the incidence of phlebitis was assessed before 72 hours after intravenous infusion.

## 3 RESULTS

### 3.1 *Phlebitis occurrence*

Table 1 shows that there were 11 patients (25.6%) experiencing phlebitis in RSI Ibnu Sina Payakumbuh during the observation period.

### 3.2 *Characteristics of infusion installation based on stab location and catheter size of IV*

Table 2 shows that the infusion location was most often in the cephalic vein (74.4%) and the most common catheter size was number 18 (46.5%).

Table 1. frequency distribution of phlebitis occurrence (n = 43).

| Variable | *f* | % |
|---|---|---|
| No phlebitis | 32 | 74.4 |
| Phlebitis | 11 | 25.6 |

Table 2. Frequency distribution by installation location and size of IV catheter (n = 43).

| Variable | *f* | % |
|---|---|---|
| a. Vena locations | | |
| Metacarpal vein | 7 | 16.3 |
| Cephalic vein | 32 | 74.4 |
| Basilic vein | 4 | 9.3 |
| b. IV catheter | | |
| Number 18 | 20 | 46.5 |
| Number 20 | 14 | 32.5 |
| Number 22 | 9 | 20.9 |

Table 3. Frequency distribution based on nurse compliance in carrying out the aseptic technique in the installation of the infusion (n = 43).

| Variable | *f* | % |
|---|---|---|
| Compliance | 19 | 44.2 |
| Non-compliance | 24 | 55.8 |

Table 4. Relationship between nurse compliance in using aseptic techniques for infusion and phlebitis in RSI Ibnu Sina Payakumbuh (n = 43).

| | Phlebitis occurrence | | | | | | |
|---|---|---|---|---|---|---|---|
| | No phlebitis | | Phlebitis | | Total | | |
| Aseptic technique | *f* | % | *f* | % | *f* | % | *p*-value |
| Compliance | 18 | 94.7 | 1 | 5.3 | 19 | 100 | 0.012 |
| Non-compliance | 14 | 58.3 | 10 | 41.7 | 24 | 100 | |

### 3.3 *Nurse compliance*

Table 3 indicates that more than half of the respondents (55.8%) did not comply with the standard aseptic techniques when performing the infusion.

### 3.4 *Bivariate analysis*

Table 4 shows that the phlebitis incidents in this study happened more frequently among nurses who were non-compliant when using aseptic techniques for the infusion (41.7%). The statistical test (chi-squared) resulted in a *p* value = 0.012 (< 0.05). This indicates that there is a significant relationship between nurse compliance when using aseptic techniques for the infusion installation and phlebitis incidence.

## 4 DISCUSSION

This study found a relatively high incidence of phlebitis (25.6%) in RSI Ibnu Sina Payakumbuh, compared with the standard ≤ 1.5% set by the Ministry of Health. This high number can damage the quality of hospital services. In addition, a significant association between a nurse's compliance when using infusion aseptic techniques and incidences of phlebitis was also identified. Forecasting researchers showed that less than half (48.8%) of nurses washed their hands before performing aseptic techniques. In fact, handwashing should be done both before and after the action (Nursalam, 2011), although the nurses also use gloves and other protective equipment. Handwashing is important in order to reduce the spread of microorganisms that are present on the hand, such that the spread of infection can be minimised and the working environment is protected from infection. In addition, all nurses can use 70% alcohol as a disinfectant. This suggests that the area to be penetrated can be disinfected with an antiseptic solution, such as providone, iodine, 70% alcohol, or chlorhexidinePerry and Potter's (2005) and did not touch the disinfected area as much as (27.9%), only 32.6% nurses did well.. A nurse ideally has a basic knowledge of the various theories related to infusion therapy. This will affect their behaviour, especially with respect to the principles relating to the prevention of complications. Nurses should be aware of the principles and techniques of asepsis, including stability, storage, labelling, interaction, dosage and calculation, and also of the appropriate equipment needed to provide safe infusion therapy to patients (Wayunah, 2012). In addition, the infusion action is delegated to the nurse, so the nurse must understand the correct methods and techniques to infuse, administer intravenous fluids and maintain intravenous systems (Potter & Perry, 2005). The incidence of phlebitis in our study was also due to more infusions being applied to the metacarpal vein (28.6%) than the other two veins. The results are in line with the studies of Lindayanti and Priyanto (2013), where the occurrence of phlebitis is based on the location of the intravenous catheter, which is most commonly placed in the distal venous region (45.5%). The location of the vein has a significant relationship with the occurrence of phlebitis because the distal blood vessels are closer to the joint and more easily moved, so that friction occurs in the vein wall due to intravenous catheters.

Potter and Perry (2005) suggested that changing the position of the limb, especially in terms of the wrist or elbow, can reduce the rate of infusion and affect blood flow. The use of a cephalic vein (located away from the wrist) is a better choice. The 6% incidence of phlebitis (the MOH standard is ≤ 1.5%) is known by observing the location of the installation and by observing the signs and symptoms of phlebitis, including pain along the cannula, erythema, redness at the stabbing site, and fever (Potter & Perry, 2005). The high incidence of phlebitis in this study is also due to the incorrect application of standard infusion procedures, such as a nurse who does not properly wash their hands, does not wear / change gloves or does not use proper disinfection techniques, and contamination of the infusion equipment during installation.

Phlebitis is an acute inflammation of the internal venous layer that is characterised by pain along the vein, redness, swelling and warmth, and can be felt around the puncture area. Phlebitis is a complication that is often associated with intravenous therapy (Nursalam, 2014Brunner&Suddarth,2013). Phlebitis can be prevented by performing aseptic techniques during infusion, using the correct IV size for the patient's veins, choosing the right veins, the type of fluid, and most importantly, the disposal of the 72-hour aseptic mounting location (Brunner & Suddarth, 2013). Our study is in accordance with the research of Rusnawati (2015), where the incidence of phlebitis in the hospital is also high (45.4%). The incidence of phlebitis is one indicator of hospital service quality. By obtaining a $p$-value of 0.012 ($< 0.05$) from the statistical test (chi-squared), it can be concluded that there is a significant correlation between compliance in using aseptic techniques for infusion installation and phlebitis occurrence. This relationship was also shown in previous research (Ince, 2010), which stated that there is a relationship between nurse compliance in applying standard infusion procedures and the occurrence of phlebitis. Phlebitis occurs due to the improper application of standard infusion procedures. The sterile principle in the installation is intended to prevent the entry of microorganisms during the installation of the infusion. According to Philip and Gorski, cited in Rusnawati (2015), aseptic techniques should be performed during every clinical procedure, including infusion, to reduce the risk of infection.

The results of our study indicate that with disinfection at the time of intravenous infusion using non-aseptic techniques, many (41.7%) have phlebitis. The study is in accordance with the statement of Phillips and Gorski, cited in Rusnawati (2015), that unsterile insertion sites are pathways for bacteria that can cause infection. This suggests that disinfection using non-aseptic techniques may increase the risk of the occurrence of phlebitis. The study also showed that of the patients who were disinfected using aseptic techniques (44.2% of the total), 5.4% also had phlebitis. The occurrence of phlebitis in patients undergoing infusion using aseptic techniques may be due to other factors, such as the physical condition of patients who move too much and often fold the infusion hand. However, large size IVs can also cause phlebitis. According to Brunner and Suddarth (2013), phlebitis can be caused by large IVs. Therefore, the size of the IV catheter should be adjusted to the conditions of the patient's blood vessel and therapy should be given.

## 5 CONCLUSION

From the description of the research results above, it can be concluded that: site setting and the selection of the IV catheter is crucial in the occurrence of phlebitis; nurse compliance in performing aseptic techniques is lacking in some aspects; there is a significant relationship between the characteristics of infusion and the incidence of phlebitis.

## REFERENCES


Ambarwati, F.R. (2014). *Konsep dasar kebutuhan manusia The basic concept of human needs.* Yogyakarta, Indonesia: Dua Satria Offset.

*Arikunto, S. (2014). Prosedur penelitian Research procedure.* Jakarta, Indonesia: Rineka Cipta.

Bruner & Suddarth, (2013). *Medical surgical nursing textbook*, 8th ed. Vol.2, Jakarta. EGC.

Cahyono, S. (2012). *Membangun budaya keselamatan pasien dalam praktik kedokteran Building a patient safety culture in medical practice.* Yogyakarta, Indonesia: Kanisius.

Darmadi. (2008). *Infeksi nosokomial problema dan pengendalianya problems of nosocomial infection and its control.* Jakarta, Indonesia: Salemba.

Depkes, R.I. (2010). *Peraturan Menteri Kesehatan Republik Indonesia Nomor 147/Menkes/Per/I/2010 tentang Perizinan Rumah Sakit Regulation of the Minister of Health of the Republic of Indonesia Number 147 / Menkes / Per / I / 2010 on Hospital Permit.* Jakarta, Indonesia: Kemenkes RI.

Gorski, L.A. (2007). Infusion nursing standards of practice. *Journal of Infusion Nursing, 30*(1), 20–21.

Hamzah, B.U. (2008). *Teori motivasi dan pengukurannya Motivation theory and measurement.* Jakarta, Indonesia: Bumi Aksara.

Hidayat, A.A. (2013). *Metode penelitian kebidanan dan teknik analisis data Method of obstetric research and data analysis techniques.* Jakarta, Indonesia: Salemba Medika.

Ignatavius, D. & Workman, M.L. (2010). *Medical-surgical nursing: Patient-centered collaborative care* (6th ed.). St. Louis, MO: Saunders Elsevier.

Ince, M, (2012). *Compliance of nurses in implementing standard operational procedures for infusion of phlebitis. Retrieved 23 August 2016 from jurnal.shb.ac.id/index.php/VM/article/download/83/83* https://www.google.co.id/?hl=en&gws_rd=ssl#hl=en&q=.

Mada D, (2012). *Nurse knowledge relation about nosocomial infection with application of sterile principle on infusion installation at Kristen Lende Moripa Hospital West Sumba. Retrieved December 22, 2016 from journal.respati.ac.id/index.php/medika/article/download/55/51* https://www.google.co.id/?hl=en&gws_rd=ssl#hl=en&q=.

Notoatmodjo. (2007). *Pendidikan dan perilaku kesehatan Education and health behavior.* Jakarta, Indonesia: Rineka Cipta.

Notoatmodjo. (2012). *Metodologi penelitian kesehatan Health research methodology.* Jakarta, Indonesia: Rineka Cipta.

Nugroho, J.S. (2008). *Perilaku konsumen Consumer behavior.* Jakarta, Indonesia: Graha Ilmu.

Nursalam. (2011). *Manajemen keperawatan Nursing management* (3rd ed.). Jakarta, Indonesia: Salemba Medika.

Nursalam. (2012). *Konsep dan Penerapan metodologi Penelitian Ilmu Keperawatan Concept and Application of Research Methodology of Nursing Science.* Jakarta, Indonesia.Salemba Medika

Nursalam. (2013). *Metodologi penelitian ilmu keperawatan Research methodology of nursing science* (3rd ed.). Jakarta, Indonesia: Salemba Medika.

Potter, P.A. & Perry, A.G. (2005). *Buku Ajar Fundamental Keperawatan: Konsep, Proses, dan Praktik Nursing Fundamentals of Nursing: Concepts, Processes, and Practices* (4th ed., Vol. 2, R. Komalasari et al., Trans.). Jakarta, Indonesia: EGC.

Putra, R. (2012). *Panduan riset keperawatan dan penulisan ilmiah Guidance on nursing research and scientific writing*. Yogyakarta, Indonesia: D—Medika.

Putri, I.R.R. (2016). Pengaruh lama pemasangan infus dengan kejadian flebitis pada pasien rawat inap di bangsal penyakit dalam dan syaraf Rumah Sakit Nur Hidayah Bantul Influence of old infusion with phlebitis incidence in inpatient at Nur Hidayah Hospital of Bantul. *Journal Ners and Midwifery Indonesia, 4*(2), 90–94.

Putri, P. (2015). *Faktor—faktor yang berhubungan dengan kepatuhan perawat dalam melaksanakan SPO tindakan suction endotracheal di RSUP Dr M. Djamil Padang Factors related to nurse compliance in the implementation of global surgical procedures of endotracheal suction measures at the hospital. M. Djamil Padang* (Unpublished undergraduate thesis, Faculty of Nursing, University of Andalas, Padang, Indonesia).

Rizky, W. (2014). Surveillance kejadian phlebitis pada pemasangan kateter intravena pada pasien rawat inap di Rumah Sakit Ar. Bunda Prabumulih *Surveillance of phlebitis incidence in intravenous catheter insertion in hospitalized patients at Ar Hospital. Bunda Prabumulih. Journal Ners and Midwifery Indonesia, 2*(1), 42–49.

Rusnawati, S. (2015). *Analisis faktor resiko terjadinya flebitis di RSUD Puri Husada Tembilahan Analysis of risk factors for the occurrence of phlebitis in hospitals Puri Husada Tembilahan* (Unpublished Master's thesis, Faculty of Nursing, University of Andalas, Padang, Indonesia).

Saam, Z. & Wahyuni, S. (2012). *Psikologi kesehatan Health psychology*. Jakarta, Indonesia: PT Raja Grafindo Persada.

Septiari, B.B. (2012). *Infeksi nosokomial Nosocomial infections*. Jakarta, Indonesia: Nusa Medika.

Smet, B. (2000). *Psikologi kesehatan Health psychology*. Jakarta, Indonesia: PT Gramedia Widiasarana Indonesia.

Susianti, M. (2008). *Keterampilan keperawatan dasar Basic nursing skills*. Jakarta, Indonesia: Erlangga.

Swanburg, R.C. (2002). *Pengantar kepemimpinan dan manajemen keperawatan Untuk perawat klinis Introduction to leadership and nursing management For clinical nurses*. Jakarta, Indonesia: EGC.

Sylvia, A. (2006). *Konsep klinis proses—proses penyakit Clinical concept of disease processes*. Jakarta, Indonesia: EGC.

Tim, I.P.C.N. (2016). Medical Record, Keperawatan, Payakumbuh *Medical Record, Nursing, Payakumbuh*: RSI Ibnu Sina. Kota Payakumbuh. Indonesia.

Ulfah. (2008). *Faktor—faktor yang terkait dengan kejadian flebitis pada pasien yang dipasang infus di bangsal RSUD Solok Factors associated with the incidence of phlebitis in patients in Solok hospital* (Thesis, PSIK, University of Andalas, Padang, Indonesia).

Wayunah. (2012). *Hubungan pengetahuan perawat tentang terapi infus dengan kejadian flebitis dan kenyamanan pasien di bangsal rawat inap RSUD Indramayu Nurse knowledge relation with phlebitis and patient comfort at Indramayu Hospital* (Thesis, Faculty of Nursing, University of Indonesia).

*Strengthening Research Capacity and Disseminating New Findings in Nursing and Public Health – Malini et al. (Eds)*


# Relationship between self-efficacy and the stress levels of B 2015 programme nursing students when completing their thesis at the faculty of nursing, Andalas University in 2016

M. Neherta, D. Novrianda & P. Raini
*Faculty of Nursing, Andalas University, Padang, West Sumatra, Indonesia*

ABSTRACT: Students from the Faculty of Nursing, Andalas University were at risk of experiencing stress because of problems in completing their final project or essay. This study aimed to determine the correlation between self-efficacy and the stress levels of the nursing students of programme B of class 2015 when completing their theses. This is a correlational study with a total sample of 66 people. This study was conducted at the Faculty of Nursing at Andalas University for eight months, from June 2016 until January 2017. Results: The result showed that the students' average self-efficacy was 92.64 and their average stress level was 15.56. Data analysis using the Pearson correlation test showed that there was a significant correlation between self-efficacy and stress levels, with moderate relation strength ($r = -0.597$) and negative direction. Conclusion: There was a significant correlation between self-efficacy and the stress levels of students at the Faculty of Nursing, Andalas University when completing their theses.

*Keywords*: Nursing Faculty, Students, self-efficacy, stress level, essay

## 1 INTRODUCTION

When entering the term essay-writing period, students begin to use their thinking ability to conduct research independently. They will use their ability to think creatively to determine the research topic, formulate problems, collect, process, and analyse data, draw conclusions from the results of the research s/he conducted, and, finally, deliver the results in the form of scientific writing (Agai-Demjaha et al., 2015).

The process of writing an essay often makes students stressed (de La Rosa-Rojas et al., 2015). Stress reactions can be positive or negative (Dugan et al., 2015). Positive stress reactions will make students more motivated to seek additional essay references and students may also become more active when working on their essay. On the contrary, negative stress reactions will cause students to avoid doing their essay or turn them towards other activities that are considered more interesting (de La Rosa-Rojas et al., 2015).

The stress that is experienced by students may either directly or indirectly affect the process of essay completion. This is because, when under stress, the individual's body will activate responses of resistance and avoidance, which will make him/her release a lot of energy that may cause fatigue both mentally and physically. Usually, this condition will be characterised by a decrease in productivity, irritability, appetite changes, difficulty in concentrating, a lower attention span, reducing the ability to remember information, and very limited decision-making ability (Dugan et al., 2015).

Due to the existence of various obstacles when completing their essay, students need a special belief. This belief that someone has about doing something, or in their ability to face obstacles, is called self-efficacy. Self-efficacy will have a great impact on stress levels and is one way to deal with difficult situations (Emamjomeh & Bahrami, 2015). Nursing students who were undertaking clinical practices concluded that self-efficacy played an important role in reducing stress and helped students to maintain positive thinking during clinical practice (Jenny et al., 2015; Sabzianpoor et al., 2015). This is also consistent with other studies that were conducted (Koreki et al., 2015; Simmonds et al., 2015), which suggested that having high self-efficacy can assist nurses in performing their duties and roles.

The Faculty of Nursing at Andalas University is one of the public nursing faculties in Indonesia that obliges students to write an academic essay as a compulsory requirement to earn a bachelor's degree. Accordingly, all students of the Faculty of Nursing at Andalas University must complete their final project in the form of an essay. At this time, the nursing students who are completing the essays are from programme B. For some of the students of programme B of class 2015, writing an essay is a new thing because, on their D3 study level, the final task was simply to make a case report or nursing care, which is very different from an essay.

The researchers' observations of the programme B nursing students of class 2015 showed that some students did not attend lectures in order to meet his/her essay supervisor, but did not get any outcome due to the supervisor being busy. The students appeared sleepy due to not sleeping until late at night in order to work on their essay. They also got frequent headaches, suffered from frequent daydreaming, and had no concentration in their subsequent lectures. Moreover, they experienced irritability, anger and appetite changes. Interviews revealed that the students felt anxious and depressed, panicked easily, were irritable, had difficulty resting, and felt less confident about finishing their essay on time. Students experienced such pressures as having difficulty meeting their supervisor, difficulty in the process of research data retrieval, difficulty in researching subjects, finding materials or references, and limited time for research. In addition to those pressures, students also had to undertake many tasks from lectures. All these pressures led them to become stressed.

Based on the phenomena described above, researchers felt a necessity to undertake research to determine the correlation between self-efficacy and the stress levels of the programme B nursing students of class 2015 when preparing an essay at the Faculty of Nursing at Andalas University in 2016.

## 2 METHOD

This was a correlational study with a total population of 66 participants from programme B of class 2015 at the Faculty of Nursing, Andalas University. Samples were taken using the total sampling technique, with a sample size of 66 people. This research was conducted at the Faculty of Nursing at Andalas University, Padang for 8 months, from June 2016 until January 2017.

This study used two types of questionnaires, which are named instruments A and B. Instrument A is a questionnaire that assesses self-efficacy, which is adopted from Nimah's (2014) questionnaire format and consists of 55 statements. The questionnaire's validity and reliability had been tested using a Cronbach's alpha value of 0.962 for 36 students of Semarang University. This questionnaire consists of 55 statements in the following composition: statements 1 to 20 for sub-variable magnitude, statements 21 to 35 for sub-variable strength, and statements 36 to 55 for sub-variable generality. The questionnaire's statements were prepared favourably (positive statements) and unfavourably (negative statements). Alternative answers to the questionnaire use a Likert scale of 1 – 4 (Very appropriate, Appropriate, Inappropriate and Very inappropriate). The evaluation criteria of the positive statements were: Very appropriate, 3; Appropriate, 2; Inappropriate, 1; Very inappropriate, 0. The negative statements were: Very appropriate, 0; Appropriate, 1; Inappropriate, 2; Very inappropriate, 3.

Instrument B is an instrument used to collect stress data, adopted from the Depression Anxiety Stress Scale (DASS) 42. DASS contains 42 item statements, which consist of 14 items on depression, 14 items on anxiety and 14 items about stress. According to Lovibond (1995),

Table 1. Frequency distribution of respondent characteristics.

| Variable | Category | Frequency | Percentage (%) |
|---|---|---|---|
| Gender | Men | 6 | 9.1 |
| | Women | 60 | 90.9 |
| | **Total** | **66** | **100.0** |
| Age | Youth end | 32 | 48.5 |
| | Early adult | 22 | 33.3 |
| | Late adult | 12 | 18.2 |
| | **Total** | **66** | **100.0** |
| Religion | Islam | 62 | 93.9 |
| | Christian | 4 | 6.1 |
| | **Total** | **66** | **100.0** |
| Marital status | Married | 31 | 47.0 |
| | Single | 35 | 53.0 |
| | **Amount** | **66** | **100.0** |

Table 2. Distribution of average value of self-efficacy and respondents' stress level.

| Variable | Mean | SD | Min–Max |
|---|---|---|---|
| Self-efficacy | 92.64 | ± 15.834 | 57–141 |
| Stress level | 15.56 | ± 5.778 | 3–36 |

Table 3. Correlation between self-efficacy and respondents' stress level.

| | Stress level | |
|---|---|---|
| Variable | R | *p* value |
| Self-efficacy | –0.597 | 0.000 |

DASS has a discriminant validity level and a reliability of 0.91, which was processed according to the Cronbach's alpha rating. For this study, we only used the part of the questionnaire about stress, consisting of 14 statement items. In order to discover the correlation, strength and direction of correlation between variables, the Pearson correlation test was applied.

## 3 RESULT

Table 1 shows that 90.9% of the respondents were female and 48.5% were in their late adolescence. Also, 93.9% of the respondents were Muslim and 53.0% were not married.

### 3.1 *Univariate analysis*

Table 2 shows that the average self-efficacy of the 66 respondents was 92.64 with a standard deviation of ± 15.834; the minimum value was 57, and the maximum value was 141. The average value of stress level for the 66 respondents was 15.56, with a standard deviation of ± 5.778, a minimum value of 3, and a maximum value of 36.

### 3.2 *Bivariate analysis*

Table 3 indicates that self-efficacy had a significant association with the respondents' level of stress when completing their essay ($p = 0.000$), with a negative correlation and correlation strength of –0.597. This means that the higher the level of self-efficacy the respondents had, the lower their stress levels became.

## 4 DISCUSSION

The results of the analysis showed $p$-value = 0.000 (< 0.05), indicating a significant correlation between self-efficacy and the stress levels of the nursing students of programme B of class 2015 who were completing their essays. A correlation value of $r = -0.597$ showed a moderate level with a negative coefficient, which means that the higher the self-efficacy that the nursing students had, the lower stress levels they faced when completing their essay. On the other hand, the lower the self-efficacy that they maintained, the higher the stress levels these nursing students experienced when completing their essays.

Therefore, self-efficacy had a direct significant negative effect on stress. This study showed that if a student had high self-efficacy, then their stress level would be lower. In other words, students having a high degree of confidence in their essay preparation resulted in a faster completion of their essay using their own ability (J.Y. Wang et al., 2015).

Another study that had previously been done discovered that the higher the self-efficacy and achievement motivation the students had, the lower stress levels they would have when doing their essay (S.M. Wang et al., 2015). This finding was supported by Ro et al. (2016), who claimed that students who had high self-efficacy would make use of all their capabilities to achieve something they desired.

Students who were preparing essays had to face certain obstacles, including difficulties in determining themes, finding references, study time limitations, repeated revision processes and delayed feedback from supervisors. Consequently, students considered essay writing to be a pressure that might cause stress (Robertson & Felicilda-Reynaldo, 2015).

The stress that students experienced during their essay preparation could lead to some negative impacts. Students avoided or delayed the writing process, they could lose concentration and motivation, and they might prefer to do other activities that were considered more interesting (Robertson & Felicilda-Reynaldo, 2015). With the existence of the various obstacles that students faced when preparing their essay, it was necessary to have a cognitive component of a belief to change the obstacles into a challenge to not just give up and to be able to face the obstacles when preparing their essay. The belief that one has in doing something or the ability to face obstacles is usually called self-efficacy (Purnamasari, 2014).

Emotional conditions, such as stress, were seen by the individual as a sign that threatened their ability. These conditions provide information about a person's level of self-efficacy and about how one can confront a task: whether they are anxious or worried (low self-efficacy) or interested (high self-efficacy). If the individual was stressed, the symptoms indicated that the individual had lower skills or abilities. Otherwise, the individual may feel that his/her self-efficacy was high if he/she had fewer symptoms of stress (Robertson & Felicilda-Reynaldo, 2015; Rogala et al., 2015), which suggests that people with high levels of stress would have a low self-efficacy.

The feelings of stress, frustration and hesitation proved that the person had a lower ability to control their emotions, while those who had a higher level of emotional control would have a higher self-efficacy (Rothberger et al., 2015). Students who had high self-efficacy would use all of their capabilities to achieve something they desired (Rowbotham & Owen, 2015). Therefore high self-efficacy will reduce the stress levels of students who are preparing an essay (Rowbotham & Owen, 2015; Rowland et al., 2015).

As professional health workers, nurses have a role as a counsellor who provides guidance or counselling. One type of counselling that nurses can give is on ways to increase self-efficacy and lower stress levels so that they do not lead to depression. With high self-efficacy, students will be able to control their stress and be capable of completing their essay according to the deadline. Therefore, it is expected that lecturers at the Faculty of Nursing, especially those who were difficult to arrange meetings with and tended to delay essay consultations, should always motivate and allocate more time for students so that students can undertake essay consultations without unnecessary delays.

## 3 CONCLUSION

In conclusion, this study discovered a significant correlation between self-efficacy and the stress levels of nursing students who were completing their essays. The correlation value of $r = -0.597$ indicated a moderate correlation with the negative coefficient. This study also recommends that lecturers at the Faculty of Nursing at Andalas University provide additional materials regarding time management and stress management, and allocate sufficient time for essay consultations with students.

## REFERENCES


Agai-Demjaha, T., Bislimovska, J.K. & Mijakoski, D. (2015). Level of work related stress among teachers in elementary schools. *Open Access Macedonian Journal of Medical Sciences*, *3*(3), 484–488.

de La Rosa-Rojas, G., Chang-Grozo, S., Delgado-Flores, L., Oliveros-Lijap, L., Murillo-Perez, D., Ortiz-Lozada, R., … Yhuri Carreazo, N. (2015). Level of stress and coping strategy in medical students compared with students of other careers. *Gaceta Medica de Mexico*, *151*(4), 443–449.

Dugan, A.M., Parrott, J.M., Redus, L., Hensler, J.G. & O'Connor, J.C. (2015). Low-level stress induces production of neuroprotective factors in wild-type but not BDNF+/– mice: Interleukin-10 and kynurenic acid. *International Journal of Neuropsychopharmacology*, *19*(3).

Emamjomeh, S.M. & Bahrami, M. (2015). Effect of a supportive-educative program in the math class for stress, anxiety, and depression in female students in the third level of junior high school: An action research. *Journal of Education and Health Promotion*, *4*, 10.

Jenny, G.J., Brauchli, R., Inauen, A., Fullemann, D., Fridrich, A. & Bauer, G.F. (2015). Process and outcome evaluation of an organizational-level stress management intervention in Switzerland. *Health Promotion International*, *30*(3), 573–585.

Koreki, A., Nakagawa, A., Abe, A., Ikeuchi, H., Okubo, J., Oguri, A., … Keio Psychiatry Resident Class. (2015). Mental health of Japanese psychiatrists: The relationship among level of occupational stress, satisfaction and depressive symptoms. *BMC Research Notes*, *8*, 96.

Nimah, Ainun. 2014. Correlation Between Social Support with Self Efficacy in Solving Nursing Student thesis at the University of Airlangga Force in 2009. *Univeristas Airlangga.*

Purnamasari, Mega Isvandiana. Correlation Self Efficacy And Achieving Motivation With The Moderate Stress Working Student Thesis. *Surakarta: University Muhammadiyah.*

Ro, Y.S., Shin, S.D., Song, K.J., Hong, S.O., Kim, Y.T. & Cho, S.I. (2016). Bystander cardiopulmonary resuscitation training experience and self-efficacy of age and gender group: A nationwide community survey. *American Journal of Emergency Medicine*, *34*(8), 1331–1337.

Robertson, D.S. & Felicilda-Reynaldo, R.F. (2015). Evaluation of graduate nursing students' information literacy self-efficacy and applied skills. *Journal of Nursing Education*, *54*(3 Suppl), 26–30.

Rogala, A., Shoji, K., Luszczynska, A., Kuna, A., Yeager, C., Benight, C.C. & Cieslak, R. (2015). From exhaustion to disengagement via self-efficacy change: Findings from two longitudinal studies among human services workers. *Frontiers in Psychology*, *6*, 2032.

Rothberger, S.M., Harris, B.S., Czech, D.R. & Melton, B. (2015). The relationship of gender and self-efficacy on social physique anxiety among college students. *International Journal of Exercise Science*, *8*(3), 234–242.

Rowbotham, M. & Owen, R.M. (2015). The effect of clinical nursing instructors on student self-efficacy. *Nurse Education in Practice*, *15*(6), 561–566.

Rowland, D.L., Adamski, B.A., Neal, C.J., Myers, A.L. & Burnett, A.L. (2015). Self-efficacy as a relevant construct in understanding sexual response and dysfunction. *Journal of Sex & Marital Therapy*, *41*(1), 60–71.

Sabzianpoor, B., Ghazanfari Amrai, M., Jalali Farahani, M., Soheila, R., Mahdavi, A. & Rahmani, S. (2015). The impact of teaching psychological welfare on marital satisfaction and self-efficacy in nurses. *Journal of Medicine and Life*, *8*(Spec Iss 4), 307–312.
Simmonds, G., Tinati, T., Barker, M. & Bishop, F.L. (2015). Measuring young women's self-efficacy for healthy eating: Initial development and validation of a new questionnaire. *Journal of Health Psychology*, *21*(11), 2503–2513.
Wang, H., Yang, X.Y., Yang, T., Cottrell, R.R., Yu, L., Feng, X. & Jiang, S. (2015). Socioeconomic inequalities and mental stress in individual and regional level: A twenty-one cities study in China. *International Journal of Equity in Health 14*, 25–31.
Wang, J.Y., Li, Y.S., Chen, J.D., Liang, W.M., Yang, T.C., Lee, Y.C. & Wang, C.W. (2015). Investigating the relationships among stressors, stress level, and mental symptoms for infertile patients: A structural equation modeling approach. *PLoS One*, *10* (10), e0140581.
Wang, S.M., Lai, C.Y., Chang, Y.Y., Huang, C.Y., Zauszniewski, J.A. & Yu, C.Y. (2015). The relationships among work stress, resourcefulness, and depression level in psychiatric nurses. *Archives of Psychiatric Nursing*, *29* (1), 64–70.

*Strengthening Research Capacity and Disseminating New Findings in Nursing and Public Health – Malini et al. (Eds)*
*© 2018 Taylor & Francis Group, London, ISBN 978-1-138-50066-2*

# Author index

9781138500662